T0137280

# Lecture Notes in Networks and Systems 760

The series "Lecture Notes in Networks and Systems" publishes the latest developments in Networks and Systems—quickly, informally and with high quality. Original research reported in proceedings and post-proceedings represents the core of LNNS.

Volumes published in LNNS embrace all aspects and subfields of, as well as new challenges in, Networks and Systems.

The series contains proceedings and edited volumes in systems and networks, spanning the areas of Cyber-Physical Systems, Autonomous Systems, Sensor Networks, Control Systems, Energy Systems, Automotive Systems, Biological Systems, Vehicular Networking and Connected Vehicles, Aerospace Systems, Automation, Manufacturing, Smart Grids, Nonlinear Systems, Power Systems, Robotics, Social Systems, Economic Systems and other. Of particular value to both the contributors and the readership are the short publication timeframe and the world-wide distribution and exposure which enable both a wide and rapid dissemination of research output.

The series covers the theory, applications, and perspectives on the state of the art and future developments relevant to systems and networks, decision making, control, complex processes and related areas, as embedded in the fields of interdisciplinary and applied sciences, engineering, computer science, physics, economics, social, and life sciences, as well as the paradigms and methodologies behind them.

Indexed by SCOPUS, INSPEC, WTI Frankfurt eG, zbMATH, SCImago.

All books published in the series are submitted for consideration in Web of Science.

For proposals from Asia please contact Aninda Bose (aninda.bose@springer.com).

Hind Zantout · Hani Ragab Hassen
Editors

# Proceedings of the International Conference on Applied Cybersecurity (ACS) 2023

*Editors*
Hind Zantout
Dubai Knowledge Park
Heriot-Watt University Dubai
Dubai, Dubai, United Arab Emirates

Hani Ragab Hassen
Dubai Knowledge Park
Heriot-Watt University Dubai
Dubai, Dubai, United Arab Emirates

ISSN 2367-3370 ISSN 2367-3389 (electronic)
Lecture Notes in Networks and Systems
ISBN 978-3-031-40597-6 ISBN 978-3-031-40598-3 (eBook)
https://doi.org/10.1007/978-3-031-40598-3



This Springer imprint is published by the registered company Springer Nature Switzerland AG
The registered company address is: Gewerbestrasse 11, 6330 Cham, Switzerland

Paper in this product is recyclable.

# Preface

We are pleased to introduce the Proceedings of The Second International Conference on Applied Cyber Security (ACS23). ACS23 aims to bring together academics, practitioners, and cybersecurity enthusiasts to discuss advances in cybersecurity and novel solutions to modern cybersecurity problems.

We continue to observe the trend of increased AI applications to tackle cybersecurity problems. Indeed, most of the papers in this volume proposed innovative ways to apply AI to address concrete cybersecurity challenges.

## Content Review

The generic topic of cybersecurity continues to evolve and expand, and this is reflected in the inclusion of two papers in this volume. 'Cyberbullying: A BERT Bi-LSTM Solution for Hate Speech Detection on Hate Speech Detection' addresses a problem that continues to blight social media. Another paper 'A cross-validated fine-tuned GPT-3 as a novel approach to fake news detection' presents a novel approach to a different problem, namely the trustworthiness of news items on the internet.

There is a clear element of cybersecurity in tackling the problem of spam, and there are good solutions already in operation. However, Arabic language spam poses particular challenges that are yet to be resolved. This is explored in 'Enhancing efficiency of Arabic Spam Filtering based on Gradient Boosting Algorithm and Manual hyperparameters tuning'.

AI is widely used in cybersecurity, and the detection of malicious activities both on hosts and on networks is one example. In 'Android Malware Detection Using Control Flow Graphs and Text Analysis', text analysis and machine learning techniques are combined for malware detection by analysis of the flow of instructions in an application. 'DroidDissector: A Static and Dynamic Analysis Tool for Android Malware Detection' describes a practical tool capable of extracting both static and dynamic features. Researchers and practitioners may find this tool particularly helpful. 'NTFA: Network Flow Aggregator' presents another tool for aggregating network traffic in a way that makes the detection of network attacks both more effective and scalable.

With considerable work underway toward an operational next-generation quantum computer, work on cybersecurity solution that can face the resulting challenges continues to make steady progress. In 'A Password-Based Mutual Authentication Protocol via Zero-Knowledge Proof Solution', a lightweight but robust protocol capable to function in that new environment is proposed.

For the convenience of readers, we divided this volume into two parts; the first reviews a panoply of AI applications to cybersecurity, whereas the second focuses on the application of AI for the detection of malicious activities.

## Target Audience

This book is suitable for cybersecurity researchers, practitioners, enthusiasts, as well as fresh graduates. It is also suitable for artificial intelligence researchers interested in exploring applications in cybersecurity. Prior exposure to basic machine learning concepts is preferable, but not necessary.

## Acknowledgements and Thanks

The editors want to express their gratitude to the authors, program committee members, and the reviewers for their great efforts writing, reviewing, and providing insightful feedback. We would also like to thank Sathya Subramaniam and Thomas Ditzinger from Springer for their help and support in the production of this manuscript.

We would like to extend our special thanks to the keynote speakers, Mr. Thomas Heuckeroth, SVP IT Infrastructure & CyberSecurity, Group CISO, The Emirates Group, UAE, and Dr. Angelika Eksteen, CEO at AIDirections, UAE.

Hind Zantout
Hani Ragab Hassen

# Conference Organization

## Honorary Chair

| | |
|---|---|
| Steve Gill | Head of School of Mathematical and Computer Sciences, Heriot-Watt University, Dubai, United Arab Emirates |

## General Chair

| | |
|---|---|
| Hind Zantout | School of Mathematical and Computer Sciences, Heriot-Watt University, Dubai, United Arab Emirates |

## General Co-chair

| | |
|---|---|
| Hani Ragab Hassen | Director of the Institute of Applied Information Security, School of Mathematical and Computer Sciences, Heriot-Watt University, Dubai, United Arab Emirates |

## Steering Committee

| | |
|---|---|
| Hani Ragab Hassen | Heriot-Watt University, Dubai, United Arab Emirates |
| Abdelmadjid Bouabdallah | University of Technology of Compiegne, France |

## Organizing Committee

| | |
|---|---|
| Hind Zantout (General Chair) | Heriot-Watt University, UAE |
| Hani Ragab (Co-chair) | Heriot-Watt University, UAE |
| Razika Sait | University of Bejaia, Algeria |
| Adrian Turcanu | Heriot-Watt University, UAE |
| Ali Muzaffar | Heriot-Watt University, UAE |

# Contents

# Malicious Activity Detection Using AI

# DroidDissector: A Static and Dynamic Analysis Tool for Android Malware Detection

Ali Muzaffar[1(✉)], Hani Ragab Hassen[1], Hind Zantout[1], and Michael A. Lones[2]

[1] Heriot-Watt University, Dubai, UAE
{am29,h.ragabhassen,h.zantout}@hw.ac.uk
[2] Heriot-Watt University, Edinburgh EH14 4AS, UK
m.lones@hw.ac.uk

**Abstract.** DroidDissector is an extraction tool for both static and dynamic features. The aim is to provide Android malware researchers and analysts with an integrated tool that can extract all of the most widely used features in Android malware detection from one location. The static analysis module extracts features from both the manifest file and the source code of the application to obtain a broad array of features that include permissions, API call graphs and opcodes. The dynamic analysis module runs on the latest version of Android and analyses the complete behaviour of an application by tracking the system calls used, network traffic generated, API calls used and log files produced by the application.

## 1 Introduction

Android has continued to evolve and gain popularity. According to statscounter [1], it holds a share of 71.74% of the smartphone market, 44.11% more than its major competitor iOS in 2023. One of the major reasons behind its popularity is the ease of access to Android applications, whether using the official Google Play Store or third party application stores.

Android's popularity brings with it a number of security concerns, malware being a major one. Malware is any software or application developed with the intention to harm the device it is installed on. The consequences of this can include monetary loss, information leakage and advertising spam. According to G Data Mobile Security Report [2] released in 2022, a new piece of Android malware is observed every 23 s. Therefore, it is essential to have a mechanism to detect malware before it causes any harm.

Traditionally, anti-malware techniques have used signature based malware detection. However, this mechanism requires maintaining a known malware database, and detecting new malware is a challenge. Recent years have seen


H. Zantout and H. Ragab Hassen (Eds.): ACS 2023, LNNS 760, pp. 3–9, 2023.
https://doi.org/10.1007/978-3-031-40598-3_1

an increase in the popularity of machine learning based approaches to malware detection, and these have shown a lot of potential [3]. Generally, these approaches require the collection of a dataset of Android applications, followed by the extraction of features using static or dynamic analysis. Our dynamic and static analysis tool (DroidDissector) extracts the most used and most useful static and dynamic features, as identified in our recent survey paper [3].

The most commonly used dynamic analysis tool in the literature is DroidBox [4] which works on Android 4.1.2, released a decade ago. The tool extracts information like SMS sent and received, read and write operations, cryptographic operations, and information leakage. The author does not maintain the tool anymore and the latest GitHub commit is over 3 years ago. AndroPyTool [5,6] extracts both static and dynamic features and stores the information in JSON files, CSV files, or writes the output to a database. The dynamic analysis module uses DroidBox and therefore only works on versions of Android up to 4.1.2. AndroPyTool also extracts system call traces in addition to the features extracted by DroidBox. Currently the most up to date and supported analysis tool is Mobile Security Framework (MobSF). The dynamic analysis module can run on both a third party virtual environment, Genymotion, or the Android virtual device which comes with Android SDK [7]. MobSF is aimed at professional environments and is not commonly used in research.

In this paper, we present DroidDissector, a fully integrated static and dynamic analysis tool for extracting features; these can then be used for Android malware detection or analysing the behaviour of an application. We developed this tool to address the limitations of currently available static and dynamic analysis tools, notably the fact that existing static analysis tools generally only extract a single type of feature (such as permissions or API calls) and that dynamic analysis tools are in most cases outdated or focus on specific commercial use cases. DroidDissector allows users to run an analysis with one central tool and can extract multiple features. The tool runs on the most up to date Android version. We used DroidDissector in two of our previous works. In [8], we carried out an investigation on API calls and how feature selection and machine learning models affect detection rates. The results showed an accuracy rate of up to 96%. We also carried out an extensive study on features for Android malware detection in [9] and used DroidDissector to extract static and dynamic features (Fig. 1).

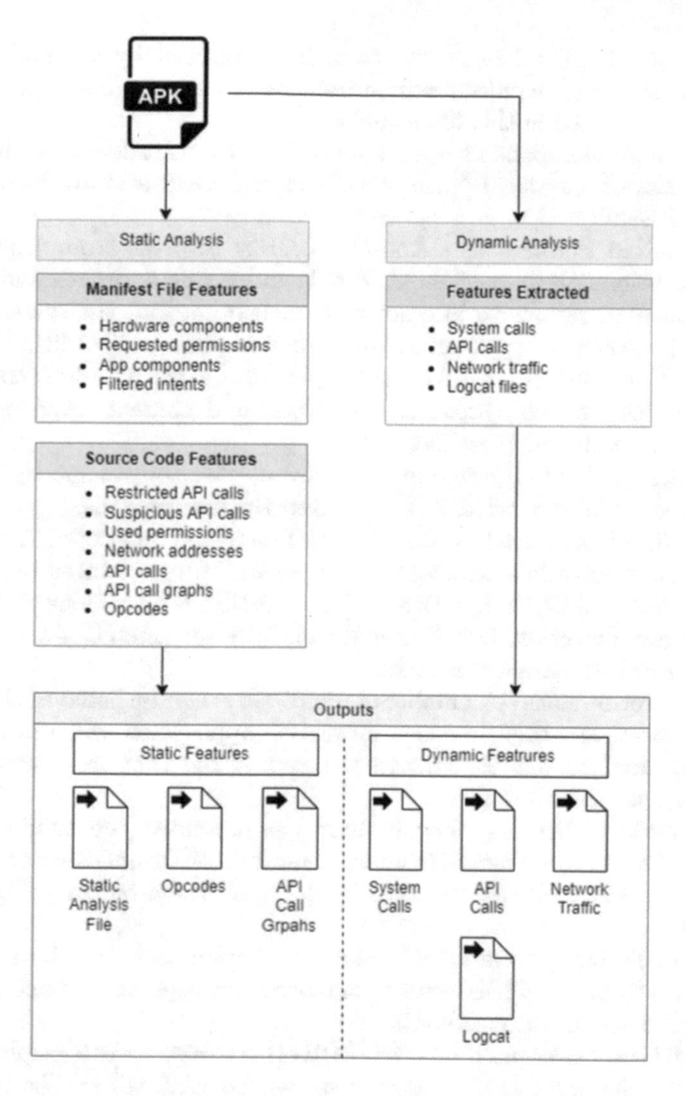


**Fig. 1.** DroidDissector Architecture

## 2 Static Analysis Tool

Static features are the most widely used features in Android malware detection. The static analysis module of DroidDissector reverse engineers and extracts the manifest file and smali files from an APK package using "APKtool" [10]. APK-Tool creates a directory for each APK file which contains the reverse engineered smali and manifest files. For each application, DroidDissector then extracts features and stores then in three separate files:

- Static Features file: This is the main file produced by the static analysis module. This file includes features from both the manifest and smali files. The features saved in this file include:
  - Hardware Components - An application can use several hardware components of the device such as camera and GPS and are defined in the manifest file.
  - Requested Permissions - Android security revolves around permissions. Permissions are part of the Android sandboxing approach and allow different applications to interact with each other and the system. Permissions used by an application are defined in the manifest file.
  - App Components - An Android application consists of four components: activities, services, broadcast receivers, and content providers and are defined in the manifest file.
  - Filtered Intents - Filtered intents allow applications to specify the intents it would like to receive, also defined in the manifest file.
  - Restricted API Calls - Drebin [11] shortlisted restricted API calls for malware detection extracted from the smali files produced by APKTool.
  - Suspicious API Calls - Drebin [11] shortlisted suspicious API calls for malware detection. DroidDissector analyzes the smali files extracted from the APK to extract this feature.
  - Used Permissions - A number of permissions can be listed in the manifest file. However, this does not mean the application will use them. The static analysis module extracts the permissions that are actually used by analysing the smali files.
  - Network Addresses - Network addresses present in the smali code.
  - API Calls v30 - The static analysis module of DroidDissector also mines for any API calls that are used by the application by analysing the smali code.
- Opcodes file: Opcodes are the low-level operations executed by the Android virtual machine. DroidDissector then stored the sequence of opcodes used by the application in an opcode file.
- API Call Graphs: We used FLOWDROID [12] to extract call graphs produced by the applications. DroidDissector stores the API call graphs in API call graphs files which can later be used to obtain the sequence of API calls which are called by the application.

The primary language used to build the static analysis module was Python 3.8. DroidDissector performs static analysis of each APK file in Python, except for the extraction of API call graphs, which uses FLOWDROID, a Java library.

## 3 Dynamic Analysis Tool

We tested the dynamic analysis tool on Linux distributions including Ubuntu and CentOS, as well as Windows operating system. DroidDissector's dynamic analysis tool requires the following to run:

- Python 3: The tool requires python 3 to run.
- Frida-Android: Frida's python library is required to hook to API calls during execution of an application.
- Android Emulator: The dynamic analysis module uses the virtual environment provided by Android SDK to run the analysis. The module runs on Android 8, 9, 10, 11, 12 and we expect it to run on Android 13.
- The tool uses *adb* from Android SDK to connect to the virtual device. The application (to be analysed) is installed using adb and inputs to the emulator are provided using adb.
- The virtual device should be running Frida-server to capture API calls called during the execution of the application.
- Static analysis report is required if API calls analysis is performed. DroidDissector requires APKTool if a report is not present to perform static analysis on the application.

### 3.1 Feature Extraction

The dynamic analysis module extracts all the major features used in Android malware detection. These include system calls, API calls, logcat files and network traffic. The module allows the user to set the number of virtual events to input to the application and the maximum time the application should run for. We use the monkey tool from Android to input virtual events to imitate normal usage of an application.

**System Calls**

The tool extracts system calls made by the application using *strace*, a Linux tool to monitor interaction between applications and the Linux kernel. It is essential to collect all the system calls made by the application to have a complete and thorough system calls analysis. We achieve this by running strace on the zygote process. This allows the tool to monitor every process running in the OS, as the Android OS uses the zygote process to start any other process. The tool filters out the system calls made by the application it is monitoring at the end of the analysis. This is different to the approach seen in past works, where researchers run the application first, and then start the system call analysis on the application, which may lead to an incomplete system calls analysis, i.e., some system calls made by the application between starting an application and getting its process ID may be lost.

Finally, the tool saves the output strace for every application in a separate file.

**Network Traffic**

Android emulator provides the user with "tcpdump" option. tcpdump is a network analyzer utility for Linux. DroidDissector uses tcpdump to analyze the network traffic of the application and stores the network traffic in "pcap" files. The pcap files can then be run on Wireshark [13] to extract and analyze network traffic of the application.

DroidDissector saves a separate network traffic for every application.

**Android Logcat**

Logcat is a command line tool provided by Android SDK that outputs the logs, system messages on errors and messages written by the programmer during the execution of the application. DroidDissector runs logcat using the adb utility and saves the log files of each application separately for further analysis.

**API Calls**

The API calls module is different to every other dynamic analysis module. Android has over 130,000 API calls, making it computationally not feasible to hook each API call. Therefore, the API calls module requires an input from the user; this could either be the static analysis report of the application stating the API calls actually used by the application in the source code, or a pre-defined list of API calls to be tracked by DroidDissector.

Frida [14] is a tool that allows hooking of API calls and monitors their execution during the run-time of an application. The tool does not miss any API call during the run-time of an application by following the process outlined. The output also includes information on the parameters passed when the application called the API. Finally, DroidDissector saves all the APIs called during execution in a separate file for further analysis.

## 4 Conclusions

The main aim of DroidDissector is to provide researchers with a powerful, centralized, and integrated tool for extracting static and dynamic features. This is especially important in static analysis, where there are very few, if any, tools available to extract multiple kinds of features from one tool. On the other hand, up to date dynamic analysis tools are not available. DroidDissector runs on the latest version of Android and is easy to set up and use for feature extraction. The list of features extracted by the tool were informed by our extensive literature review. Our aim is to keep improving this tool, adding functionalities to extract more features and integrate more tools.

## References

1. StatCounter: Mobile operating system market share worldwide (2023). Accessed 16 Jan 2023. http://gs.statcounter.com/os-market-share/mobile/worldwide
2. G Data: G data mobile security report: conflict in Ukraine causes decline in malicious android apps (2022)
3. Muzaffar, A., Hassen, H.R., Lones, M.A., Zantout, H.: An in-depth review of machine learning based android malware detection. Comput. Secur. 102833 (2022)
4. Droidbox (2014). https://github.com/pjlantz/droidbox
5. Martín García, A., Lara-Cabrera, R., Camacho, D.: Android malware detection through hybrid features fusion and ensemble classifiers: the andropytool framework and the omnidroid dataset. Inf. Fusion **52**, 128–142 (2018)

6. Martín, A., Lara-Cabrera, R., Camacho, D.: A new tool for static and dynamic android malware analysis. In: Data Science and Knowledge Engineering for Sensing Decision Support: Proceedings of the 13th International FLINS Conference (FLINS 2018), pp. 509–516. World Scientific (2018)
7. Mobile security framework (MOBSF). https://github.com/MobSF/Mobile-Security-Framework-MobSF. Accessed 10 Jan 2023
8. Muzaffar, A., Ragab Hassen, H., Lones, M.A., Zantout, H.: Android malware detection using API calls: a comparison of feature selection and machine learning models. In: Ragab Hassen, H., Batatia, H. (eds.) ACS 2021, pp. 3–12. Springer, Cham (2021). https://doi.org/10.1007/978-3-030-95918-0_1
9. Muzaffar, A., Hassen, H.R., Zantout, H., Lones, M.A.: A comprehensive investigation of feature and model importance in android malware detection (2023). https://arxiv.org/abs/2301.12778
10. R. Connor Tumbleson. Apktool (2019). https://ibotpeaches.github.io/Apktool/
11. Arp, D., Spreitzenbarth, M., Hübner, M., Gascon, H., Rieck, K.: DREBIN: effective and explainable detection of android malware in your pocket. In: Network and Distributed System Security Symposium (NDSS) (2014)
12. Arzt, S., Huber, S., Rasthofer, S., Bodden, E.: Denial-of-app attack: inhibiting the installation of android apps on stock phones. In: Proceedings of the ACM Conference on Computer and Communications Security, pp. 21–26 (2014)
13. Wireshark. https://www.wireshark.org/. Accessed 10 Jan 2023
14. Frida: A world-class dynamic instrumentation toolkit. https://frida.re/. Accessed 10 Jan 2023

# Android Malware Detection Using Control Flow Graphs and Text Analysis

Ali Muzaffar(✉), Ahmed Hamza Riaz, and Hani Ragab Hassen

Heriot-Watt University, Dubai, UAE
{am29,ar2015,h.ragabhassen}@hw.ac.uk

**Abstract.** The Android OS has a massive user-base with hundreds of millions of devices. Due to the growth of this platform, an increasing number of malicious applications are becoming available to download online. We propose a tool that, when provided a sample application, performs binary classification of the sample as either malicious or benign. We constructed control flow graphs from the API and library calls made by the sample application. We then used control flow graphs to train classification models that utilized text analysis methods such as TF-IDF. Our technique reported accuracy rates of up to 95%.

**Keywords:** Android malware · Control Flow graph · API calls

## 1 Introduction

The Android OS ecosystem has been on a steady rise in popularity since the launch of Android 1.0 in 2008. In 2023, 71.74% of mobile devices run Android, which amounts to nearly 2.5 billion devices as of 2019 [1]. Nearly 40% of these devices are running a version older than Android Oreo (API 25), despite the latest version being Android 13 (API 33). These older versions of Android may no longer supported by Google and have critical security flaws. The detection of malware, therefore, becomes critically important especially when the sheer number of devices at risk is considered. As per the experimentation conducted by Dey et al., about 2%, or close to 40,000 applications of the applications on the Google Play Store were flagged as malicious [2].

The Google Play Store is managed by Google, who also manage the Android operating system - as a result, users may have a false sense of security while downloading and installing applications. Maliciously repackaged applications are also widespread on the Google Play Store and other popular APK hosting websites.

Our aim is to develop a scalable model that can perform binary classification on Android applications. The classification model will make use of text analysis techniques applied on decompiled Android applications. To the best of our knowledge, this is the first time CFGs are used with TF-IDF on bigram API calls for Android malware detection.

The rest of this paper is organized as follows: Sect. 2 covers background information along with a review of research done previously, Sect. 3 presents a detailed


H. Zantout and H. Ragab Hassen (Eds.): ACS 2023, LNNS 760, pp. 10–20, 2023.
https://doi.org/10.1007/978-3-031-40598-3_2

review of the dataset used and the feature extraction process, Sect. 4 discusses the results of the experiments and compares our results with other approaches and finally Sect. 5 concludes the paper and discusses possible future work.

## 2 Related Work

There are three main methods of classifying the malicious nature of an app using API calls. The first is the consideration of the sequences of API calls extracted from DEX files, obtained from decompiled application packages, as done in Maldozer [3]. The API sequences are extracted, discretized and used to train a convolutional neural network which is later used for recognition tasks. In place of call sequences, subsequences of system calls can also be used to train both machine and deep learning models, as demonstrated by Sabhadiya et al., and Lin et al. [3,4]. Adding the analysis of intents, broadcast receivers and services, along with requested permissions obtained from AndroidManifest.xml (from decompiled packages) has also shown to be proficient at classification of malware and malware families.

The second method of detection of malicious application in the Android ecosystem is via the use of control flow graphs (CFGs). With this method, each function call and API call made by the application is logged and the sequence obtained is converted into a CFG. The graphs produced by applications are very characteristic of their behavior - function names and code can be obfuscated, but the pattern of calling functions and APIs manifests itself very clearly in the flow of control. Similarities in the graphs and subgraphs can then be compared to create an efficient classifier. Previous works have shown very promising results when using CFGs to detect malware [5].

Lastly, the code of the malicious applications themselves can also be analyzed. As per Taheri, Kadir, and Lashkari [6], n-gram analysis of sequential API calls by an application is very effective in the binary classification of malware. The performance of such classifiers can be further improved with the inclusion of network flows. In our previous work [7], we used API calls extracted from the source code of the applications and reported accuracies of up to 96%.

Lin et al. noticed that the ever-growing popularity of the Android ecosystem has resulted in the ease of creation and distribution of maliciously repackaged applications (MRAs) [4]. Indeed, these MRAs are found in several places including the official Google Play Store. Detection of these application therefore falls under two main categories: static analysis and dynamic analysis. Static analysis, while easy to execute, fails in most cases due to the nature of code obfuscation and general changes in the structure of the code. The authors proposed a tool based on the dynamic analysis of applications, named **SCSDroid**: a tool that analyzes the System Call Sequences whenever they are made to the Android operating system. They claim that even if the code is obfuscated and changed, the application cannot hide its malicious behavior from the system calls that it produces.

Jaiswal et al. [8] propose an automatic detection of MRAs that works by the dynamic analysis of system calls. The permissions and system calls, along with their quantity and frequency are analyzed to understand the application's behavior. The Linux tool **strace** is used to extract a system call summary that includes the application ID, system call function, frequency and duration of the call. From their experiments, they observe that certain system calls such as "clock_gettime" or "futex" occurred more than other calls in benign applications, and that calls such as bind, brk and getsocket occurred with higher frequency in malicious games.

Jiang et al. analyzed the opcode sequences generated by applications, as these provide rich semantic and structural information about the malware [9]. Nine classification models are used by the authors, and they are: k-nearest neighbors, linear support vector machine, RBF-support vector machine, decision trees, random forests, AdaBoost, logistic regression, gradient boosting classifier, and multi-layer perceptron Classifier. They implemented these in Python and used a dataset of 5560 samples of malware from 178 families. 2-fold cross validation is carried out from a dataset constructed from the top 40 families. They achieve 99% accuracy and AUC with k-nearest neighbors, Decision trees and Gradient boosting classifier, but the training time for Gradient Boosting is significantly higher than with k-nearest neighbors.

The authors then conclude by pointing out that code written with advanced obfuscation techniques and cryptographic methods may result in sensitive opcode sequences not being captured. Additionally, as they conduct supervised learning, only previously known classes families of malware can be used as classifiers and day-zero attacks with new types of malwares would render their methods ineffective.

Taheri et al. [6] presented a two-layer malware analyzer that incorporated n-gram sequential relations in the network flows. They started by making a comparison of various datasets and evaluated them based on 15 criteria. From these datasets they then selected the ones best fit for their application and started creating their two-layer detection tool. The first layer they made is the Static Binary Classification (SBC) layer which learns to differentiate between benign and malicious applications. This works by analyzing the permissions and intents as present in the AndroidManifest.xml, from which 8115 features were extracted and used in the random forest algorithm. The dataset had a 60–40 split for training and testing.

The applications that were classified as malware from the SBC layer are sent into the second layer, called the Dynamic Malware Classification (DMC) layer. This layer's aim was to classify a given malware into a type (Adware, Ransomware, Scareware, SMSware) and a family (from 39 families). This dynamic analysis component made use of API calls and network activity. The NLTK python toolkit was utilized to conduct an n-gram analysis of the API calls, and from their experimentation they found that 2-gram analysis gives more accurate results. The malware binary classification resulted in a 95.2% recall and a 81.5% recall in the malware category classification, and a 61.2% recall in the malware family classification.

Zhang et al. [10] proposed a framework that used text classification to detect malicious applications. In their methodology, the application's .apk file was first decompiled and then four sets of features were extracted from the decompiled files. These are: permissions, services, intents, and receivers. These features were used as input vectors in a TextCNN model with five layers: embedding, convolutional, max-pooling, fully connected, and finally, softmax. Each input to their model was presented as a $n \times k$ feature map, where "each sequence is a k-dimensional vector (padded where necessary), and each input is present as a series of n sequences".

They conducted four experiments with 12,254 benign applications crawled from Anzhi (a Chinese app store) and 11,912 malicious applications (from the Contagio Community and the Android Malware Genome Project). Benign applications were tagged as benign only after several virus checkers such as VirusTotal did not detect the application as malicious. They trained their TextCNN model with a learning rate of 1e-2 for 10 epochs and used a batch size of 16, and each dataset had a train-test ratio of 90–10. They used K-nearest neighbor, random forests, logistic regression, Naïve Bayes (TF-IDF), and FastText text classification algorithm to evaluate the performance of their model. Out of datasets D1, D2 and D3, their framework almost always achieves an accuracy greater than 96%, and achieves better accuracy, F-score, and recall when compared to other classification algorithms.

Liu et al. [11] proposed a feature generation technique by extracting relevant function calls, broadcasts, and permissions from decompiled Android applications. They consider three elements of the application: sensitive broadcasts, sensitive permissions and finally, bigrams of API calls. When functionA is followed by functionB, this is considered to be a bigram and functionA and functionB are analyzed as one unit.

Before these features are extracted, they build a CFG for the application and then filter out irrelevant nodes and edges. This process of applying a graph reduction transformation results in a CFG that is significantly smaller graphs. After constructing these transformed CFGs, they analyze 1,000 Android applications and picked the 300 most common tuples of function calls (bigrams). In addition, they use TF-IDF to select the 20 most popular sensitive permissions and broadcasts to finally built a feature set of 345 features. They then used a dataset of 4972 samples and train five classifiers with a train/test split of 3732/1240. The five models they tested with were the Naive Bayes, random forest, logistic regression, support vector machines and k-nearest neighbor. Their best performer was the random forest Classifier which achieved 95.4% accuracy and 96.5% recall.

The main issue with their methodology is that they did not mention how they picked their most popular bigrams of functions from the malicious APKs. If the most frequently-occurring function calls were considered, then an additional set of tests should have been conducted where some sort of normalization was also done.

Zhu et al. [12] aimed to automate the generation of malicious behavior of an application as a sequence of API calls. Their dataset consisted of 600 malicious applications from 30 malware families and 600 additional benign application from the Google Play Store. The dataset was split into five sets and 5-fold cross validation was carried out to ensure the viability of the results. The training samples are then vectorized by comparing if the input features are present in the sample application's CFG. These vectors are then used to build models with the support vector machine, Naïve Bayes and ID3 classifiers. From their testing phase, ID3 achieves a 91.5% AUC, Naïve Bayes achieves 87.2% and support vector machine achieves 92.4% AUC. The authors also discussed the effect of varying the minimum sequence length and the support, and found that the best results are achieved with 90% support and minimum sequence length of 2.

## 3 Framework

In this section, we discuss the dataset we created and used, as well as the feature extraction process.

### 3.1 Dataset

Our dataset contains 5,500 benign Android applications and 5,863 malicious Android applications. We did verify our Android applications (APK files) using VirusTotal [13]. SHA-256 hashes of each application were uploaded to the VirusTotal servers via their API (v3), and we then used the returned malware analysis reports to label our dateset.

**Table 1.** Dataset malware family distribution

| Malware Family | Count |
|---|---|
| Trojan | 1,981 |
| Adware | 682 |
| Potentially Unwanted Application (PUA) | 524 |
| Virus | 200 |
| None | 2,714 |

Table 1 shows the distribution of malware families in this dataset. There are a large number of applications tagged as None - this is due to very few (1–3) security vendors flagging them as malicious. The dataset was collected from VirusShare [14].

Due to processing issues with FlowDroid [15], and additional unicode errors due to the obfuscated code in the extracted CFGs, certain applications could not be used to train or test the classifiers. The final number of applications was: 4,998 benign applications and 5,446 malicious applications. These were split into training and testing sets, with 4532/475 applications in the benign train/test set and 4,984/462 applications in the malware train/test set.

### 3.2 Data Extraction and Preprocessing

This section discusses the various steps required to extract the relevant data from a dataset of Android applications. This extracted data is then processed so that it can be used to train various classification models.

FlowDroid [15] is a taint analysis tool that is used to detect injection vulnerabilities in Android applications. A part of FlowDroid's functionality is the extraction of CFGs. These CFGs are multigraphs that represent all the possible entry- and exit-points of an application.

For example, consider the following code:

```
functionX();
functionY();
functionZ();
```

FlowDroid converts this to a graph with two edges:

```
functionX() functionY()
functionY() functionZ()
```

Any line in the graph represents an edge. The first item in the edge is the source function and the second item is the destination function. The more complex an application is, the larger the resulting CFG is. From the dataset used in this project, the largest CFG occupied 875MB of disk space and had 1.8 million edges. Around 600GB of CFGs were extracted from the dataset in total.

Figure 1 demonstrates the iteration through a CFG produced by FlowDroid. Each edge has three elements: a context function, a source function, and a destination function. For the purposes of this project, only the source and destination function are used.

Figure 2 is an example of a few edges from a CFG as they were output by FlowDroid.

From Fig. 2, it is evident that there is a lot of extra information in the output CFG from FlowDroid. The input to the classifiers must be a graph where each line contains the source function name, followed by a separator, followed by the destination function. To achieve this, we wrote a parser to convert it into a simplified edges as shown in Fig. 3.

```
for (Iterator<Edge> edgeIt = Scene.v().getCallGraph().iterator(); edgeIt.hasNext(); )
{
    Edge edge = edgeIt.next();

    SootMethod smSrc = edge.src();
    Unit uSrc = edge.srcStmt();
    SootMethod smDest = edge.tgt();

    fw.write( str: uSrc + ", " + smSrc + ", " + smDest + "\n");
    fw.flush();
}
```

**Fig. 1.** Iterating through the CFG produced by FlowDroid

```
62  specialinvoke $r0.<java.lang.RuntimeException: void <init>(java.lang.String)>("Stub!"), <java.lang.Object: void finalize()>, <java.lang.
    RuntimeException: void <init>(java.lang.String)>
63  virtualinvoke $r1.<android.media.MediaPlayer: void stop()>(), <com.prof21.aaanjah.MainActivity: void stopPlying()>, <android.media.MediaPlayer:
    void stop()>
64  virtualinvoke $r1.<android.media.MediaPlayer: void release()>(), <com.prof21.aaanjah.MainActivity: void stopPlying()>, <android.media.
    MediaPlayer: void release()>
65  specialinvoke r0.<java.lang.Exception: void <init>()>(), <java.lang.RuntimeException: void <init>(java.lang.String)>, <java.lang.Exception: void
    <init>()>
66  specialinvoke $r1.<java.lang.RuntimeException: void <init>(java.lang.String)>("Stub!"), <java.lang.RuntimeException: void <init>(java.lang.
    String)>, <java.lang.RuntimeException: void <init>(java.lang.String)>
67  specialinvoke r0.<java.lang.Throwable: void <init>()>(), <java.lang.Exception: void <init>()>, <java.lang.Throwable: void <init>()>
68  specialinvoke $r1.<java.lang.RuntimeException: void <init>(java.lang.String)>("Stub!"), <java.lang.Exception: void <init>()>, <java.lang.
    RuntimeException: void <init>(java.lang.String)>
69  specialinvoke r0.<java.lang.Object: void <init>()>(), <java.lang.Throwable: void <init>()>, <java.lang.Object: void <init>()>
70  specialinvoke $r1.<java.lang.RuntimeException: void <init>(java.lang.String)>("Stub!"), <java.lang.Throwable: void <init>()>, <java.lang.
    RuntimeException: void <init>(java.lang.String)>
```

**Fig. 2.** A few edges from a CFG

```
dummyMainMethod|dummyMainMethod_com_unity_androidnotifications_UnityNotificationManager
dummyMainMethod|dummyMainMethod_com_unity3d_player_UnityPlayerActivity
dummyMainMethod_com_unity3d_player_UnityPlayerActivity|<clinit>
dummyMainMethod_com_unity3d_player_UnityPlayerActivity|<clinit>
<clinit>|<clinit>
<clinit>|Log
<clinit>|loadLibrary
<clinit>|<clinit>
Log|<clinit>
Log|w
Log|e
dummyMainMethod_com_unity_androidnotifications_UnityNotificationManager|<init>
dummyMainMethod_com_unity_androidnotifications_UnityNotificationManager|onReceive
onReceive|<clinit>
onReceive|sendNotification
```

**Fig. 3.** Edges in a simplified CFG

### Term Frequency-Inverse Document Frequency

Term frequency-inverse document frequency is a statistical technique that determines the importance of a word in a document. For this project, the importance of an edge in a CFG. Applying the TfidfVectorizer from scikit-learn [16] with no limit on the number of features results in a vectorizer with 1.2 million features. Through experimentation, we decided to set the parameter max_features to 5,000.

## 4 Experimental Results

This section discusses the results of the various models that were trained on the dataset. We trained five classifiers, namely Naive Bayes, support vector machines, logistic regression, random forests and dense network. Each model was trained using 5-fold cross validation and we used grid search to find the optimal hyper-parameters.

```
Model: "sequential"
_________________________________________________________________
Layer (type)                 Output Shape              Param #
=================================================================
dense (Dense)                (None, 128)               640128
_________________________________________________________________
dropout (Dropout)            (None, 128)               0
_________________________________________________________________
dense_1 (Dense)              (None, 64)                8256
_________________________________________________________________
dropout_1 (Dropout)          (None, 64)                0
_________________________________________________________________
dense_2 (Dense)              (None, 32)                2080
_________________________________________________________________
dropout_2 (Dropout)          (None, 32)                0
_________________________________________________________________
dense_3 (Dense)              (None, 1)                 33
=================================================================
Total params: 650,497
Trainable params: 650,497
Non-trainable params: 0
_________________________________________________________________
```

**Fig. 4.** Dense network we trained with 3 layers

Figure 4 shows the setup for the dense network. Each dense layer is followed by a dropout layer, and it was trained with 32 epochs. Binary crossentropy was used as the loss function and Adam was used as the optimizer.

For Naive Bayes, after running GridSearchCV, the best parameter was $alpha = 0.001$. For SVM, the best parameter was $C = 50$. Logistic regression model was trained with $C = 10$ and $penalty = l1$. A grid search on the random forest model showed the best parameters were $criterion = entropy$ and $n\_estimators = 200$ (Table 2).

Table 3 shows that the best performer is the dense network, followed by random forest. Random forests trained with 200 trees and linear SVC took the longest time to train compared other models. This gives the dense network an edge over the random forest model not only in the evaluation metrics but also in training time.

The models discussed above were tested with the top 300 features which were obtained by scikit-learn's CountVectorizer.

**Table 2.** Detailed evaluation results

| Model | Accuracy | Precision | Recall |
|---|---|---|---|
| Naive Bayes | 0.91 | 0.91 | 0.91 |
| Linear SVC | 0.93 | 0.93 | 0.93 |
| Logistic Regression | 0.93 | 0.93 | 0.93 |
| Random Forest | 0.94 | 0.94 | 0.94 |
| Dense Network | 0.95 | 0.95 | 0.95 |

**Table 3.** Top 300 features evaluation results

| Model | Accuracy | Precision | Recall |
|---|---|---|---|
| Naive Bayes | 0.69 | 0.85 | 0.83 |
| Linear SVC | 0.74 | 0.75 | 0.74 |
| Logistic Regression | 0.83 | 0.85 | 0.83 |
| Random Forest | 0.93 | 0.94 | 0.93 |
| Dense Network | 0.66 | 0.75 | 0.66 |

Table 3 shows that apart from the random forest, the other models do not perform particularly well. The fact that CFG reduction was not carried out and that sensitive broadcasts and permissions were not utilized must be kept in mind. Random Forest performs well due to its capability of both handling a large number of features and its ability to generalize over a dataset.

Liu et al. [11] used CFG in their work, proposing a graph reduction technique to reduce the size of their graph and use 300 graph features. They also used TF-IDF to select the most relevant permissions and broadcasts from their dataset. In contrast, we only used CFGs to select top 300 features via TF-IDF. Our emulation results demonstrate that by just using TF-IDF applied on CFG, we can achieve similar detection rates with random forest as Lui et al., who reported 95% accuracy and 96.5% recall.

Furthermore, while using the dense network on the full feature set of 5,000 features, dense network produced an accuracy of 95% accuracy and recall of 95%, which are very close to the results obtained by Lui et al.

## 5 Conclusion

The rapid growth of the Android OS has led to a significant increase in the number of malicious applications available online. The objective of this paper was twofold: firstly, to create a tool to make a simplified representation of the CFG and secondly, we aimed to deliver a model that could conduct binary classification of malware. We developed five different models that achieved this goal and were able to get an accuracy of above 91%.

Our findings suggest that relying solely on popular edges (300) from a set of CFGs is not a viable approach for building an Android malware classifier except from using random forest. However, when using 5,000 features, all the models reported an accuracy of above 91%. Future work can explore improvements to our methodology such as extracting CFGs from obfuscated applications, using TF-IDF on other features beyond CFG and improving better feature reduction.

## References


1. StatCounter: Mobile operating system market share worldwide (2023). http://gs.statcounter.com/os-market-share/mobile/worldwide. Accessed 16 Feb 2023
2. Dey, A., Beheshti, L., Sido, M.-K.: Health state of Google's playstore - finding malware in large sets of applications from the android market. In: ICISSP (2018)
3. Sabhadiya, S., Barad, J., Gheewala, J.: Android malware detection using deep learning. In: 2019 3rd International Conference on Trends in Electronics and Informatics (ICOEI), pp. 1254–1260 (2019)
4. Lin, Y.-D., Lai, Y.-C., Chen, C.-H., Tsai, H.-C.: Identifying android malicious repackaged applications by thread-grained system call sequences. Comput. Secur. **39**, 340–350 (2013). https://www.sciencedirect.com/science/article/pii/S0167404813001272
5. Muzaffar, A., Hassen, H.R., Lones, M.A., Zantout, H.: An in-depth review of machine learning based android malware detection. Comput. Secur. 102833 (2022)
6. Taheri, L., Kadir, A.F.A., Lashkari, A.H.: Extensible android malware detection and family classification using network-flows and API-calls. In: 2019 International Carnahan Conference on Security Technology (ICCST), pp. 1–8 (2019)
7. Muzaffar, A., Hassen, H., Lones, M.A., Zantout, H.: Android malware detection using API calls: a comparison of feature selection and machine learning models. In: Ragab Hassen, H., Batatia, H. (eds.) ACS 2021, pp. 3–12. Springer, Cham (2021). https://doi.org/10.1007/978-3-030-95918-0_1
8. Jaiswal, M., Malik, Y., Jaafar, F.: Android gaming malware detection using system call analysis. In: 2018 6th International Symposium on Digital Forensic and Security (ISDFS), pp. 1–5 (2018)
9. Jiang, J., et al.: Android malware family classification based on sensitive opcode sequence. In: 2019 IEEE Symposium on Computers and Communications (ISCC), pp. 1–7 (2019)
10. Zhang, N., Tan, Y.A., Yang, C., Li, Y.: Deep learning feature exploration for android malware detection. Appl. Soft Comput. **102**, 107069 (2021). https://www.sciencedirect.com/science/article/pii/S1568494620310073
11. Liu, X., Lei, Q., Liu, K.: A graph-based feature generation approach in android malware detection with machine learning techniques. Math. Probl. Eng. **2020**, 1–15 (2020)
12. Zhu, J., Wu, Z., Guan, Z., Chen, Z.: API sequences based malware detection for android. In: 2015 IEEE 12th International Conference on Ubiquitous Intelligence and Computing and 2015 IEEE 12th International Conference on Autonomic and Trusted Computing and 2015 IEEE 15th International Conference on Scalable Computing and Communications and Its Associated Workshops (UIC-ATC-ScalCom), pp. 673–676 (2015)
13. Virustotal. https://www.virustotal.com
14. Virusshare. https://virusshare.com. Accessed 10 Jan 2023

15. Arzt, S., et al.: Flowdroid. ACM SIGPLAN Not. **49**, 259–269 (2014)
16. Buitinck, L., et al.: API design for machine learning software: experiences from the scikit-learn project. In: ECML PKDD Workshop: Languages for Data Mining and Machine Learning, pp. 108–122 (2013)

# NTFA: Network Flow Aggregator

Kayvan Karim(✉), Hani Ragab Hassen, and Hadj Batatia

Heriot -Watt University, Edinburgh, Scotland
{k.karim,H.RagabHassen,H.Batatia}@hw.ac.uk

**Abstract.** Network intrusion detection systems (NIDS) play a vital role in defending against cybersecurity threats. One effective way of detecting attacks is to analyse their footprint on the network traffic logs. Flow-based logging is a standard method for logging network traffic. Given the high volume of traffic, it is unpractical to manually analyse it. This is where machine learning can play a great role by automatically analysing traffic logs and identifying attacks. One way to process the flow data is to aggregate the information and extract insight from the aggregated network traffic. This paper presents an open-source aggregator NTFA (Network Flow Aggregator). This customisable network flow aggregator can aggregate flow-based, from the standard NetFlow format, data based on the time as well as other criteria, while offering the operator the option of using different time windows. To evaluate the suitability of the output aggreates for the intrusion detection task, we use this tool to aggregate the CIDDS-001 dataset and train a classifier using a decision tree algorithm. The model can classify aggregated network traffic data with an accuracy of 99.8%. Our experiment demonstrates that the aggregated data can be used in various machine-learning research projects or industries related to intrusion detection scenarios.

**Keywords:** Network Intrusion Detection · Netflow · CIDDS

## 1 Introduction

With the advent of high-speed networks, traditional intrusion detection techniques that rely on analyzing the complete payloads of packets or protocol headers using packet-based or protocol-based inspection have faced challenges and might become obsolete. This is due to the significant computational resources required to perform such inspections, making them impractical for today's networks [2]. Flow-based network traffic logging is a technique used to capture and analyze network traffic flow within a network. This approach involves collecting data on the individual packets of data that pass through a network and grouping their summaries into larger entities known as flows. A flow is defined as a packet sequence that shares common characteristics, such as source and destination addresses, protocol types, and port numbers [3].

Flow-based intrusion detection uses features from network flows to profile benign and malicious traffic patterns, offering lower computational and time


H. Zantout and H. Ragab Hassen (Eds.): ACS 2023, LNNS 760, pp. 21–28, 2023.
https://doi.org/10.1007/978-3-031-40598-3_3

complexity than packet-based or protocol-based inspection, since a network flow is smaller in size than the packets from which the flow was extracted. This results in a faster process that does not require payload analysis. However, due to the information lost when converting network packets into a flow, it can be difficult for NIDS to efficiently distinguish malicious flows from benign ones [2].

Aggregated flows-based inspection is an extended version of flow-based inspection that analyzes a host's behaviour in a computer network by combining multiple flows with the same IP address. This method is more effective than single-flow-based inspection, as it captures the characteristics of attacks involving multiple flows and amplifies their statistical features, leading to better intrusion detection [4].

In this study, we propose NTFA (Network Flow Aggregator), a new tool that can be used as an initial pre-processing step to aggregate the flow data and provide more insight into the behaviour of each source IP in a time frame. The user of our tool can set any time frame, and this tool can process any flow-based data using dynamic column mapping. Dynamic column mapping is a simple method to map any NetFlow data column to the required feature column in the aggregator.

The rest of this paper is organised as follows; Sect. 2 reviews notable relevant work. Section 3 explains our methodology. We evaluate our aggregator in Sect. 4 by applying a machine learning model on its output to classify traffic. Section 5 concludes our paper.

## 2 Related Work

Some research has been done to pre-process the flow-based data like NetFlow and convert the categorical data into numerical ones. Ring et al. proposed IP2Vec [5], a vector-based representation of IP addresses, inspired by Word2Vec [6]. The main goal of IP2Vec is to create a vector space representation of IP addresses that can be used for various tasks in network security, such as anomaly detection, classification, and clustering. The IP2Vec approach uses the context of an IP address to create a vector representation. In other words, the context in which an IP address is used can provide valuable information about the nature and behaviour of the IP address. For example, an IP address that is frequently used to access certain types of websites or services might be more likely to be associated with a particular type of behaviour or activity. However, this method uses only one flow and not the information provided from the aggregation of the flows.

Moustafa et al. utilised Simple Random Sampling (SRS) and Association Rule Mining (ARM) techniques to build a module for a NIDS [7]. The authors used SRS for selecting relevant observations, illustrating its substantial impact on IDS design. The study results showed that the SRS-based sampling technique proves to be highly efficient for summarisation purposes, as it significantly reduces the processing time required for data analysis. However, the ARM approach was superior to the sampling technique for detection tasks. This advantage can be attributed to the fact that, unlike sampling, ARM retains all flow information, ensuring no crucial data is lost during the detection process.

Wu et al. proposed a new approach to identifying malicious activity in the cybersecurity industry and on the internet has been suggested by researchers in a recent study [4]. The paper develops a Latent Semantic Analysis Framework for Intrusion Detection (LSAFID), which uses K-means and PLSA to convert numerical network flow features into a semantic-based topic model. The authors used aggregation with a window of one minute to amplify subtle features and reduce network flow numbers. In conclusion, these experiments demonstrated that aggregation of flow-based data could be an essential step in detecting intrusion, and having a variation in the time frame window can provide more flexibility for aggregation.

## 3 Methodology

The present study introduces the Network Flow Aggregator, NTFA, a tool designed to aggregate and analyse flow-based network data within a specified time frame. The tool was developed to enhance the aggregation process of flow-based data by providing a simple and easy-to-use tool for researchers and industry experts. This tool is available as an open-source solution on GitHub.

NTFA is a Python-based tool without any dependency on a third-party library. The main script uses a JSON configuration file to process the flow-based data. In this file, the user stores the index of different columns that NTFA requires to process the data. The user fills in the configuration information during the initial process of the data. Then, the configuration file gets stored for further use. After identifying the required column, the user can set the processing frame size. The frame size is a time window in seconds. During the data flow process, the aggregator uses this time frame window as the base for the aggregation frame.

The aggregation happens in each time frame. The aggregator takes each source IP in the frame and aggregates the flow parts. The features and the aggregation process are listed in Table 1. The aggregation process has two phases.

**Table 1.** Aggregated features list

| Feature Name | Aggregation Process |
|---|---|
| Start Frame | DateTime ISO formatted string of the start time of the frame |
| End Frame | DateTime ISO formatted string of the start time of the frame |
| Source IP | Source IP is unique in each frame |
| Duration | total duration of the flow |
| Protocol | Count of unique protocols |
| Source Port | Count of unique Source Port |
| Destination IP | Count of unique Destination IP |
| Destination Port | Count of unique Destination Port |
| Number Of Packets | total number of packets in the flow |
| Label | First spotted label |

**Table 2.** Sample Aggregated Output Data - 3600-second frame

| Feature Name | Value |
|---|---|
| start frame | 2017-03-22T00:00:00.071000 |
| end frame | 2017-03-22T01:00:00.071000 |
| src ip | 192.168.220.14 |
| duration | 0.178 |
| protocol | 2 |
| src port | 4 |
| dst ip | 3 |
| dst port | 4 |
| number packets | 138 |
| label | normal |

First, the time frame information is built for each source IP. In this frame, the duration and number of packets in the same source IP flows are added up and stored in a separate column. Each unique protocol, source port, destination IP, and destination port is identified and added to separate lists. If the flow data has a label, the unique labels for the frame flow are also collected and stored in a list. In the second Phase, the final output of the aggregator is processed. Output data is based on each source IP in each time frame: the duration and number of packets are added up, and for the protocol, source port, destination IP, and destination port, the number of unique values is counted and stored in the output. The first spotted NetFlow label is used for the aggregate label. A sample of one frame of output data can be seen in Table 2.

## 4 Evaluation

Our main goal is to produce an aggregation tool. We wanted to see if the aggregated output still has enough information in it to build machine learning models with reasonable performance. So. we used the output data in a simple classification algorithm to evaluate the aggregation result and check if the aggregation process could keep the network behaviour. First, we aggregated the CIDDS-001 data [1], and to classify the aggregated data, we utilised a decision tree model. Specifically, we aggregated the week two data of the OpenStack dataset from Coburg Intrusion Detection Data Sets CIDDS-001 over a 3600-second time frame. This dataset contained 10,310,735 flow data records over one week.

After processing the data using our aggregator tool, we obtained 98,148 aggregated records. We then used these records to train a decision tree model with Gini impurity and without a maximum depth. To evaluate the model's performance, we used 10-fold cross-validation and calculated the mean accuracy over the ten folds. We obtained an average accuracy of 99.8%, while working on an aggregate whose size is around 1% of the original data size.

Additionally, we analysed the aggregated data's features. For example, we observed a linear correlation between the number of source ports and the number of destination IP addresses and observed a similar linear correlation between duration and the number of destination IP addresses, as shown in Fig. 1. We also created a histogram of the numerical data to better understand its distribution, as shown in Fig. 2.

Overall, our evaluation of the aggregator tool demonstrated its ability to process large amounts of flow data and reduce the number of records. Also, the output data amplifies the network's behaviour and provides meaningful numerical values for the machine learning algorithms. The results also highlight the usefulness of decision tree models in classifying aggregated data and identifying patterns.

The aggregated data contained the labels shown in the list below. The attacker and victim are clearly identified in the labels.

**Labels in Aggregated Data**

- normal
- portScan victim
- portScan attacker
- DOS victim
- DOS attacker
- pingScan victim
- pingScan attacker
- bruteForce victim
- bruteForce attacker

In this study, we aimed to develop a flow-based network data aggregator called NTFA. The experiment with the dataset demonstrated that NTFA reduced the data size from 10,310,735 to 98,148, which is approximately a 99% reduction in the data size and supplies meaningful numerical values for the machine learning algorithms. We used the output to train a decision tree model and used ten-fold cross-validation to ensure its reliability. The mean accuracy of the model was calculated over the ten folds, and we found it to be 99.8%.

The model we designed can classify traffic as either benign or malicious and further classify the malicious traffic into its precise type of attack. Therefore, we believe that our aggregator can be a valuable tool for network security professionals and researchers in detecting and mitigating potential attacks. Overall, our study's results demonstrate the effectiveness of NTFA as a flow aggregator and highlight the importance of accurate traffic classification and attack detection in network security.

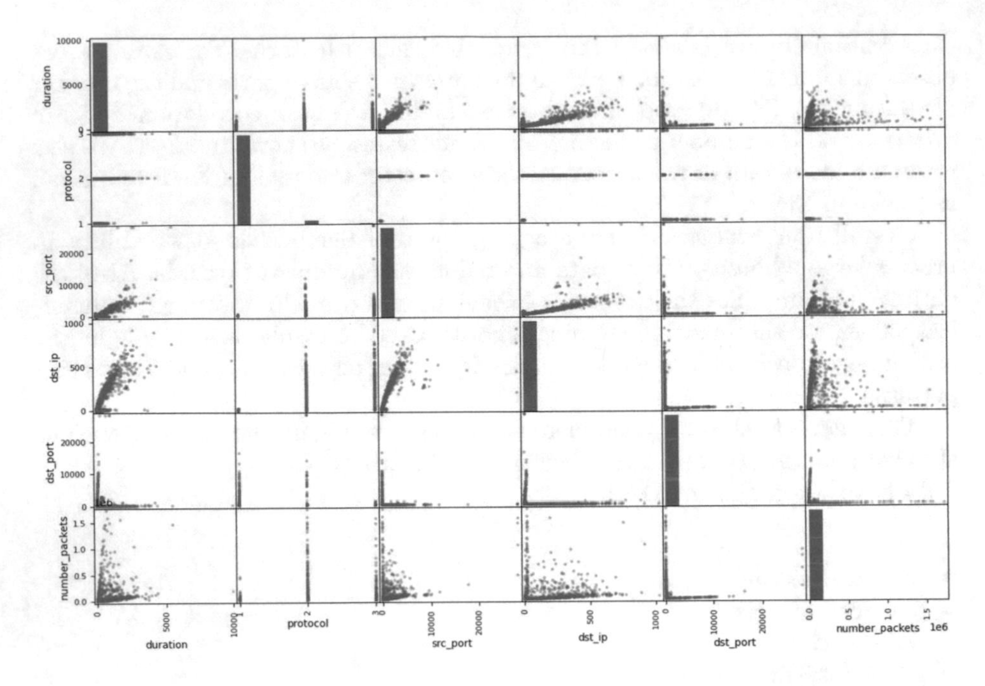


**Fig. 1.** The co-relation between features of the aggregated data

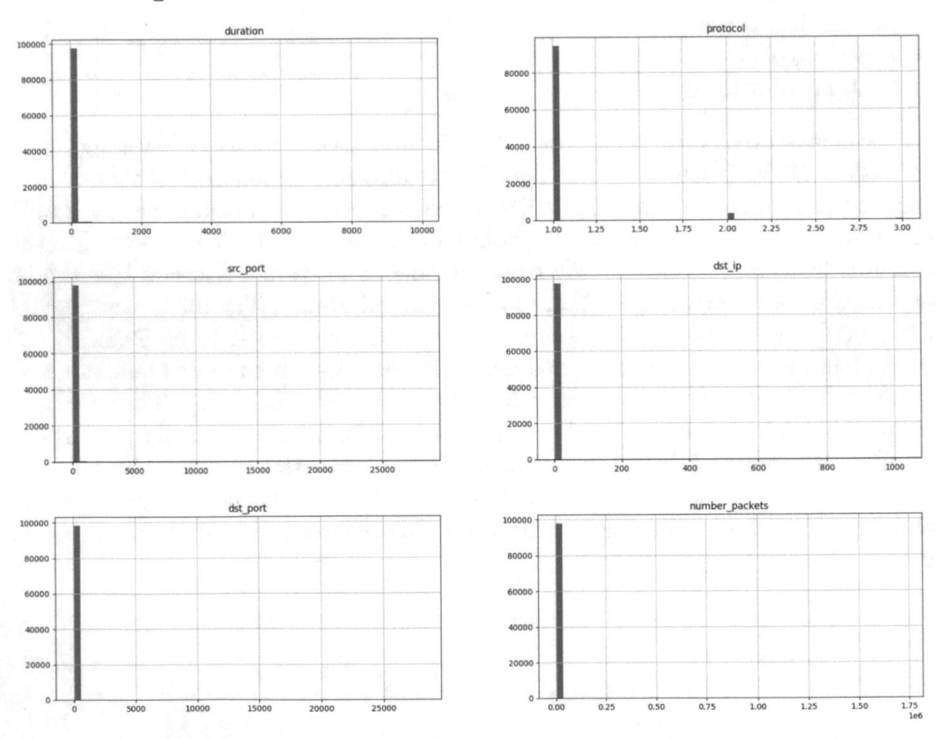


**Fig. 2.** Demonstrate the histogram of all the numerical values in the processed data.

## 5 Conclusion

The paper describes a tool that can comprehensively aggregate flow data within a specific time frame window to analyse the flow data's behaviour. The tool's primary objective is to help network administrators better understand their network traffic patterns and identify any potential security threats. While the tool has proven to be helpful, there are some limitations in its current implementation.

One of the limitations is that the aggregator can include more features like the total number of bytes transferred, which can provide more comprehensive insights into the flow data's behaviour. Future plans include incorporating additional features and evaluating different time frame windows to identify the optimal frame size. This will help ensure that the tool can handle large data volumes and provide accurate and reliable results.

By incorporating additional features and optimising the time frame window size, the tool's functionality and efficiency can be improved and used to analyse complex network traffic patterns accurately. Ultimately, the tool's improved functionality will help network administrators better understand their network traffic and detect potential security threats promptly and effectively.

## References

1. Ring, M., Wunderlich, S., Gruedl, D., Landes, D., Hotho, A.: Flow-based benchmark data sets for intrusion detection. In: Proceedings of the 16th European Conference on Cyber Warfare and Security (ECCWS), pp. 361-369. ACPI (2017)
2. Jadidi Z, Muthukkumarasamy V, Sithirasenan E, Sheikhan M. Flow-based anomaly detection using neural network optimized with GSA algorithm. In2013 IEEE 33rd International Conference on Distributed Computing Systems Workshops, pp. 76-81. IEEE, 8 July 2013
3. Ring, M., Schlör, D., Landes, D., Hotho, A.: Flow-based network traffic generation using generative adversarial networks. Comput. Secur. **82**, 156–172 (2018)
4. Wu, J., Wang, W., Huang, L., Zhang, F.: Intrusion detection technique based on flow aggregation and latent semantic analysis. Appl. Soft Comput. **127**, 109375 (2022)
5. Ring, M., Dallmann, A., Landes, D., Hotho, A.: IP2Vec: learning similarities between IP addresses.' In: 2017 IEEE International Conference on Data Mining Workshops, ICDMW, vol. 2017, pp. 657-666 (2017)
6. Mikolov, T., et al.: Distributed representations of words and phrases and their compositionality. In: Advances in Neural Information Processing Systems (2013). arXiv : 1310 . 4546v1 [ cs . CL ] 16 Oct 2013
7. Moustafa, N., Creech, G., Slay, J.: Flow aggregator module for analysing network traffic. In: Pattnaik, P.K., Rautaray, S.S., Das, H., Nayak, J. (eds.) Progress in Computing, Analytics and Networking. AISC, vol. 710, pp. 19–29. Springer, Singapore (2018). https://doi.org/10.1007/978-981-10-7871-2_3
8. Bhuyan, M.H., Bhattacharyya, D.K., Kalita, J.K.: Network Traffic Anomaly Detection and Prevention. CCN, Springer, Cham (2017). https://doi.org/10.1007/978-3-319-65188-0

9. M. Ring, S. Wunderlich, D. Grüdl, D. Landes, and A. Hotho, "Flow-based benchmark data sets for intrusion detection. In: Proceedings of the 16th European Conference on Cyber Warfare and Security, ECCWS, vol. 16, pp. 361-369 (2017)
10. Claise B, Sadasivan G, Valluri V, Djernaes M. Cisco systems netflow services export version 9 (2004)
11. Tran, Q.A., Jiang, F., Hu, J.: A real-time netflow-based intrusion detection system with improved BBNN and high-frequency field programmable gate arrays. In:2012 IEEE 11th International Conference on Trust, Security and Privacy in Computing and Communications, pp. 201-208. IEEE, 25 June 2012

# AI Applications in Cybersecurity

# A Password-Based Mutual Authentication Protocol via Zero-Knowledge Proof Solution

Mostefa Kara[1](✉), Konstantinos Karampidis[2], Zaoui Sayah[1], Abdelkader Laouid[1], Giorgos Papadourakis[2], and Mohammad Nadir Abid[1]

[1] LIAP Laboratory, University of El Oued, PO Box 789, 39000 El Oued, Algeria
{mostefakara,abdelkader-laouid}@univ-eloued.dz

[2] Department of Electrical and Computer Engineering, Hellenic Mediterranean University, Heraklion, Crete, Greece
{karampidis,papadour}@hmu.gr

**Abstract.** Password-based authentication is the most common strategy despite it having considerable problems. Password can be stealthily observed, its management is costly where users have to change their passwords regularly. With the appearance of Quantum Computing, the classic authentication models are threatened to be hacked. In this work, we propose a secure authentication scheme based on the zero-knowledge model. In the proposed scheme, the verifier generates random numbers $r_1$ and $r_2$ and mixes them with the stored password; after extracting $r_1$, the user proves the possession of its password using $r_1$ and another random number $r_3$. Therefore, both the prover and verifier authenticate each other with only a single message, which provides a lightweight protocol with robust resistance against known attacks. The proposal can resist quantum computer attacks if we use a post-quantum key-establishment algorithm to exchange the password for one-time.

**Keywords:** Authentication · Security · Anonymity · Privacy · Post-Quantum

## 1 Introduction

In the modern era, the interconnection of devices requires user authentication. Although new authentication schemes have been proposed [1], they are vulnerable to attacks [2] and password-based authentication remains the most utilized authentication scheme. However, major drawbacks like weak or replicated passwords can be reported. Moreover, people tend to forget passwords and therefore new ones must be generated through a secure procedure. This raises major security issues. Malicious users try to brute-force passwords to gain access to services or devices.

Let $m$ be the number of accepted characters and $n$ be the required password length ($m > n$). The complexity -in terms of Big O notation- of a brute


H. Zantout and H. Ragab Hassen (Eds.): ACS 2023, LNNS 760, pp. 31–40, 2023.
https://doi.org/10.1007/978-3-031-40598-3_4

force password cracker would be $O(m^n)$. For instance, if the number of accepted characters (uppercase letters, lowercase letters, numbers, symbols, etc.) is 96 and the password consists of 8 characters, the complexity of a brute force password algorithm would be $O(96^8)$, which is translated to 6,634,204,312,890,625 possibilities. This number may seem very big, however, the time needed from a modern computer to crack it would be no more than 2 h. It is obvious that as the number of used characters is increasing, the complexity and therefore the time needed would increase respectively. That is if we use a 10-character password the time needed to crack it, would rise to 2 years. This may seem a solution, but we must consider that computers with more computational power appear, like quantum computers.

Quantum computing offers great computational power and makes use of the quantum states of subatomic particles - the so-called qubits- to store information. Quantum computers are believed to solve problems that -from a feasible time aspect- are inevitable for classical computers. Indeed, in 2019 a Google quantum computer performed calculations that would have taken thousands of years on an ordinary computer [3]. One of these complex calculations could be password cracking. Furthermore, besides the risk of cracking passwords, privacy must be considered as well. Authentication schemes must deploy mechanisms in which neither of the two parties must reveal sensitive information to the other.

Thus, the need of proposing new authentication techniques that are robust to attacks and respect privacy is crucial. In this paper, we propose a novel authentication scheme based on the zero-knowledge model. It is a method by which one party (the prover) can prove to another party (the verifier) that a given statement is accurate while the prover avoids sending any information - e.g., the password- apart from the affirmative answer that the statement is correct. In this authentication protocol, there is no exchange of passwords and thus the communication is secure, and privacy is respected.

There will be no spoofing of transmitted data. Only legitimate users will gain a chance to access the remote host (server), and the user has to submit his username (ID) and password (PW). The proposed protocol maintains a zero SMS policy with no additional charges for SMS. It encrypts and secures data inside the server from any misuse. In this work, we will focus to illustrate the Authentication Phase (including Login Phase).

The rest of the paper is as follows. In Sect. 2, we discussed the related works. The model architecture is described in detail in Sect. 3. Section 4 performs an analysis of the proposed scheme. Finally, the conclusion is presented in Sect. 5.

## 2 Related Work

In the literature, many authentication protocols for different purposes have been proposed. To improve security in electronic payments, authors in [4] proposed an approach focused on the Time-based OTP authentication algorithm with biometric fingerprints, whoever the algorithm uses a secret key is switched between the client and the server. The shuffle of the TOTP method improved by screening the key as being a QR code. The main problem with this technique it is designed only for electronic payment authentication with many restrictions.

To warrant the security of data communication between mobile devices, authors in [5] developed a prototype based on a smooth projective hash function (SPHF) and commitment-based password-hashing schemes (PHS) to construct an asymmetric PAKE protocol protected against quantum attacks. While this asymmetric-PAKE protocol insures security and efficiency by using the Bellare-Pointcheval-Rogaway (BPR) model. The main drawback of the proposed scheme is its use of either a message authentication code (MAC) or a one-time signature [6], which involves additional time and cost.

In [7], the authors proposed a protocol of authentication and key agreement to guarantee the anonymity of a mobile user in wireless mobile environments. Furthermore, they used a new method of computing the temporary identity (TID) during the authentication process which is calculated by a user at the beginning and updated by both user and network edge during the protocol execution. The major issue with this proposal is the use of public key cryptosystems based on the Discrete Logarithm Problem, which means it is not safe against quantum attacks.

In [8], the authors proposed a dynamic password-based solution adapted for a wireless sensor network environment and access control problem. This approach provides a very light computational load and requires simple operations, by using a one-way hash function and exclusive-OR operations. The authors here did not discuss the resistance against quantum computers.

Paper [9] proposed an authentication procedure for third-generation mobile systems where the authentication process is made as a protocol suite involving two subprotocols: a certificate-based authentication (CBA) protocol and a ticket-based authentication (TBA) protocol. The authentication procedure uses both public- and secret-key cryptosystems. In this scheme, the parameter $R_1$ was sent in clear text, but only the client and server can get $R_2$, session key $K$, $K = R_1 + R_2$. In a quantum attack, the adversary can get $R_2$ and $k$.

In [10], for lightweight IoT devices and a large number of scenarios, authors proposed a decentralized authentication and access control mechanism. The public blockchain [11] and fog computing are used in this mechanism which provides great results against blockchain-based authentication techniques. The device encrypts the password with its private key and sends it to the blockchain-enabled fog node to decrypt it using the device's public key. The main drawback here is the large size of the messages.

To avoid these vulnerabilities, we propose a new password-based authentication scheme using lightweight computation. We will prove that the proposed authentication protocol is more resistant to various attacks than other related schemes.

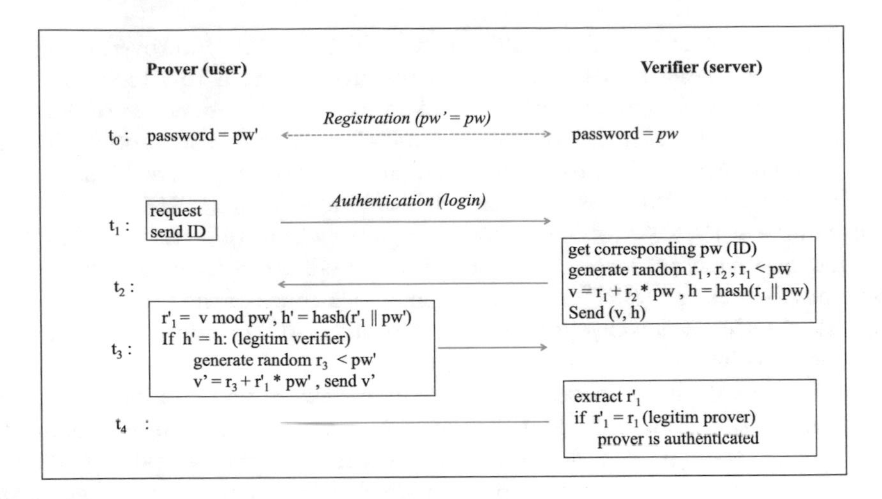


**Fig. 1.** Proposed password authentication scheme

## 3 Proposed Technique

In the proposed protocol (Fig. 1), it is assumed that the user (prover) and the remote host (verifier) have previously the same passwords $pw'$ and $pw$ respectively, which means that we will not discuss the registration phase. This password is exchanged for the first time in a secure channel and for one time, after that the authentication process does not need to send PW every time the user wants to log in. Rather, random numbers are exchanged within the zero-knowledge proof protocol, where the user proves to the remote host that he knows the password without sending any secret information to it.

The proposed password-authenticated key exchange PAKE model can be achieved if the user and the server, who hold the same password, prove to each other that they know the password without disclosing any useful information about it. In the end, they establish a shared secret session key, assuring security against (offline) password-guessing attacks.

Our scheme ensures that both the prover and the verifier end up generating the same strong session key if both know the same password, and independent session keys otherwise.

When the user makes an authentication request after entering his password $pw'$, the server generates two random numbers $r_1$ and $r_2$ with $r_1 < pw$ where $pw$ is the password stored in the server; The first random number is used for verification and it is considered a challenge, the second one is used to hide the password. Then, the server calculates a new value $v$ which is defined by Eq. 1, and sends $v$ and $h$ to the user, where $h$ is defined by Eq. 2.

$$v = r_1 + r_2 \times pw \tag{1}$$

$$h = hash(r_1 \parallel pw) \tag{2}$$

After receiving $(v, h)$, and extracting $r_1'$ by using Eq. 3, the user system (e.g. smartcard) computes $h'$ which is equal to $h' = hash(r_1' \parallel pw')$ then verifies if $h' = h$.

$$r_1' = v \mod pw' \tag{3}$$

We note here that any other party that does not have the same password cannot calculate the correct value of $r_1'$.

Then, the user generates another random number $r_3$ as a noise to hide $r_1' \times pw'$, where $r_3 < pw'$, then calculates $v'$ (Eq. 4) and sends it to the server.

$$v' = r_3 + r_1' \times pw' \tag{4}$$

The server must recover $r_1'$ using $pw$ by computing $r_3 = v' \mod pw$; then $r_1' = (v' - r_3) \div pw$. Finally, the server checks if the calculated $r_1'$ is equal to the sent $r_1$; if same, the user is authenticated; because this means that the requester has the same password.

The authentication process can be shown in Algorithms 1 and 2.

**Algorithm 1** Verifier algorithm

**Require:** requester ID
**Ensure:** authentication 0 : 1

1: **function** VERF
2: get corresponding $pw$ of ID
3: generate random $r_1, r_2$ with $r_1 < pw$
4: compute $v \leftarrow r_1 + r_2 \times pw$
5: compute $h \leftarrow$ hash$(r_1 \parallel pw)$
6: send $v, h$
7: wait until receive $v'$
8: compute $r_3 \leftarrow v' \mod pw$
9: compute $r_1' \leftarrow (v' - r_3) \div pw$
10: **if** $r_1' = r_1$ **then**
11: establish key
12: **else**
13: disconnect
14: **end if**
15: **end function**

**Algorithm 2** Prover algorithm

**Require:** Password $pw'$

**function** PROOF
2: send request (ID)
wait until receive $v$, $h$
4: compute $r_1' \leftarrow v \mod pw'$
compute $h' \leftarrow \text{hash}(r_1' \parallel pw')$
6: **if** $h' = h$ **then**
generate random $r_3$ with $r_3 < pw'$
8: compute $v' \leftarrow r_3 + r_1' \times pw'$
send $v'$
10: **else**
disconnect
12: **end if**
**end function**

## 4 Scheme Analysis and Performance

We present a Password Authenticated Key Exchange (PAKE) system. The essential reason is that PAKE can limit the influence of any password leakage. In particular, PAKE allows a remote host to send the password hash, which guarantees that the enemy has to use an offline dictionary attack to recover it from hashed passwords. A zero-knowledge proof is introduced by the user to convince the remote host that he has knowledge of the password for every login attempt. Also, the remote host requests to show the client that he knows the right credential for the registered password.

In the registration phase, it is necessary to establish the password one time between the prover and the verifier. Without considering quantum computing, the current modified Diffie-Hellman key exchange protocol might be sufficient for now. But an attack against our model can be on the initial key establishment itself using quantum computing. A quantum computer efficiently solves the discrete logarithm problem for both finite fields and elliptic curves. Hence, algorithms like DH and RSA are also at risk. Therefore, to pass our protocol to a quantum-safe authentication, we need to use a post-quantum key exchange technique such as Classic McEliece, CRYSTALS-KYBER, NTRU, and SABER.[1]

In classic schemes, a dictionary attack consists of intercepting the hash of the password and trying all dictionary words. In our scheme, this type of attack is more difficult because the intercepted hash is vary in each authentication process ($h = hash(r_1 \parallel pw)$ with $r_1$ is random in each process). Therefore, an attacker needs two dictionary attack successes with the aim to compare between $(r_1 \parallel pw)$ and $(r_1' \parallel pw)$ and extract $pw$.

[1] https://csrc.nist.gov/Projects/post-quantum-cryptography/post-quantum-cryptography-standardization/round-3-submissions.

Man-in-the-middle (MITM) attacks are when an enemy wishes to eavesdrop on or intercept a message and alter the intended message for the recipient. Users and servers receive messages without verifying the sender, without realizing an enemy has inserted themselves between them. Mutual authentication can prevent MITM attacks because both the user and server verify each other using the same password. To realize MITM, an adversary has to know (or guess) the random value $r_3$ that is into intercepted message $v'$ (Eq. 4); at the same time, has to know (or guess) the random value $r_1$ that is into intercepted message $v$ (Eq. 1). Therefore, if one of the parties is not verified who is the sender, the session will end.

A replay attack is like a MITM attack in which older messages are replayed out of context to fool the server. However, this does not work against techniques using mutual authentication. We can add timestamps because they are a verification factor. If the change in time is greater than the maximum allowed time delay in the protocol, the session will be aborted. Also, messages can use random numbers to keep track of when a message was sent.

Since the adversary doesn't have access not only to passwords, but also to random values $r_1$, $r_2$, $r'_1$, and $r_3$. This is implicitly proved as a result of the fact that in the CCA game, an adversary enables to query to encryption and decryption oracles, but has no access to underlying random values. Therefore even a quantum adversary does not have the ability to distinguish passwords from random values.

Spoofing attacks are based on using false information to pose as another user in order to gain access to the remote host. Mutual authentication like in our protocol can prevent this type of attack because the remote host will authenticate the user as well, and verify if he is a legitimate user or not before allowing any further communication and access. This verification is done by the possession of the correct password via zero-knowledge proof. Even if an attacker somehow gets the user ID, the attacker will still need the password to spoof the user's identity.

In impersonation attacks when the user and server authenticate each other, they send each other a hidden password that only the other party can exploit, verifying themselves as a trusted source. In this way, attackers cannot use impersonation attacks because they do not have the correct password to act as if they are the other party. If a malicious third party wants to fake a user, he can only guess the password; the success probability is one of $2^\lambda$ where $\lambda$ is the password length.

Our mutual authentication model supports zero-trust networking because it can protect communications against attacks. It also guarantees information integrity [12] because if the parties are verified to be the correct source, then the received data is reliable as well.

We ensure data privacy; when the user and server create a random session key, they protect the messages communicating between them; hence, an adversary gets nothing from the radio channel. Also, we ensure perfect forward secrecy;

**Table 1.** A comparative summary of computation (ms) and communication (bit) costs

| Technique | Computation | Communication |
|---|---|---|
| Ref. [15], 2019 | 80.163 | 3456 |
| Ref. [16], 2019 | 80.032 | 2624 |
| Ref. [17], 2020 | 3 | 1600 |
| Ref. [18], 2021 | 80.6 | 1600 |
| Ref. [19], 2022 | 9.158 | 5844 |
| Ours | 0.2 | 576 |

even if a session key is disclosed, due to the freshness of $r_1$, $r_2$, $r'_1$, and $r_3$ in each session, an adversary cannot compute old or new session keys.

The proposal is suitable for a large network where clock synchronization is difficult [13,14].

Table 1 demonstrates the experimental result which is made on PC-i3. In the implementation, we used Hash-256 and a password size equal to 80 bits. As shown in Table 1, our scheme has the lowest computation and communication cost compared with other schemes. The very small computation time is due to the use of very few arithmetic operations including two multiplications and just only one execution of the hash function. For example, [15] uses four multiplications and five hash function executions. Our scheme contains a single message from prover to verifier which consists of $(v, h)$, $v$ equal to $r_1 + r_2 \times pw$, if $r_1$, $r_2$, and $pw$ are 80 bits in size, $v$ will be 160 bits in size, with $\mid h \mid = 250$ bits. The prover sends also a single message $v'$ of size 160 bits; so we have a total of 576 bits.

## 5 Conclusion

The goal of post-quantum cryptography is to provide cryptographic models that are secure against quantum machines and at the same time can interoperate with existing communications protocols. In this article, we introduced a new lightweight authentication method, which is explained easily and clearly. The advantage of this technique is that it relies on a zero-knowledge proof that is considered quantum-resistant cryptography by including post-quantum Key-establishment Algorithms. The proposed protocol is free of cost, and it works charging no extra cost. Analysis proved the solidity of our proposed method and that it is a practical technique. We achieved 0.2 ms in computation cost and 576 bits in communication cost. In the future, we hope to make an extended version that gives more details about the proposed protocol, including the registration phase.

## References


1. Karampidis, K., Linardos, E., Kavallieratou, E.: StegoPass – utilization of steganography to produce a novel unbreakable biometric based password authentication scheme. In: Gude Prego, J.J., de la Puerta, J.G., García Bringas, P., Quintián, H., Corchado, E. (eds.) CISIS - ICEUTE 2021. AISC, vol. 1400, pp. 146–155. Springer, Cham (2022). https://doi.org/10.1007/978-3-030-87872-6_15
2. Karampidis, K., Rousouliotis, M., Linardos, E., Kavallieratou, E.: A comprehensive survey of fingerprint presentation attack detection. J. Surveill. Secur. Saf. **2**(4), 117–161 (2021)
3. Gibney, E.: Hello quantum world! google publishes landmark quantum supremacy claim. Nature **574**(7779), 461–463 (2019)
4. Hassan, A., Shukur, Z., Hasan, M.K.: An improved time-based one time password authentication framework for electronic payments. Int. J. Adv. Comput. Sci. Appl **11**(11), 359–366 (2020)
5. Li, Z., Wang, D., Morais, E.: Quantum-safe round-optimal password authentication for mobile devices. IEEE Trans. Dependable Secure Comput. **19**(3), 1885–1899 (2020)
6. Kara, M., Laouid, A., Hammoudeh, M.: An efficient multi-signature scheme for blockchain, Cryptology ePrint Archive (2023)
7. Go, J., Park, J., Kim, K.: Wireless authentication protocol preserving user anonymity. In: WPMC'01. WPMC, pp. 1153–1158 (2001)
8. Wong, K.H., Zheng, Y., Cao, J., Wang, S.: A dynamic user authentication scheme for wireless sensor networks. In: IEEE International Conference on Sensor Networks, Ubiquitous, and Trustworthy Computing (SUTC 2006), vol. 1, p. 8. IEEE (2006)
9. Tzeng, Z.-J., Tzeng, W.-G.: Authentication of mobile users in third generation mobile systems. Wireless Pers. Commun. **16**, 35–50 (2001)
10. Khalid, U., Asim, M., Baker, T., Hung, P.C., Tariq, M.A., Rafferty, L.: A decentralized lightweight blockchain-based authentication mechanism for IoT systems. Cluster Comput. **23**(3), 2067–2087 (2020)
11. Kara, M., et al.: A novel delegated proof of work consensus protocol. In: 2021 International Conference on Artificial Intelligence for Cyber Security Systems and Privacy (AI-CSP), pp. 1–7. IEEE (2021)
12. Kara, M., Laouid, A., Hammoudeh, M., Bounceur, A., et al.: One digit checksum for data integrity verification of cloud-executed homomorphic encryption operations, Cryptology ePrint Archive (2023)
13. Kara, M., Laouid, A., Bounceur, A., Hammoudeh, M.: Secure clock synchronization protocol in wireless sensor networks, prePrint (2023)
14. Kara, M.: A lightweight clock synchronization technique for wireless sensor networks: a randomization-based secure approach, prePrint (2023)
15. Wang, J., Wu, L., Choo, K.-K.R., He, D.: Blockchain-based anonymous authentication with key management for smart grid edge computing infrastructure. IEEE Trans. Ind. Inf. **16**(3), 1984–1992 (2019)
16. Kaur, K., Garg, S., Kaddoum, G., Guizani, M., Jayakody, D.N.K.: A lightweight and privacy-preserving authentication protocol for mobile edge computing. In: 2019 IEEE Global Communications Conference (GLOBECOM), pp. 1–6. IEEE (2019)
17. Alladi, T., Bansal, G., Chamola, V., Guizani, M., et al.: Secauthuav: a novel authentication scheme for UAV-ground station and UAV-UAV communication. IEEE Trans. Veh. Technol. 69(12), 15-068–15-077 (2020)

18. Sadhukhan, D., Ray, S., Biswas, G., Khan, M.K., Dasgupta, M.: A lightweight remote user authentication scheme for IoT communication using elliptic curve cryptography. J. Supercomput. **77**, 1114–1151 (2021)
19. Wang, C., Wang, D., Xu, G., He, D.: Efficient privacy-preserving user authentication scheme with forward secrecy for industry 4.0. Sci. China Inf. Sci. **65**(1), 112301 (2022)

# A Cross-Validated Fine-Tuned GPT-3 as a Novel Approach to Fake News Detection

Karim Hemina(✉), Fatima Boumahdi, Amina Madani, and Mohamed Abdelkarim Remmide

LRDSI laboratory, sciences faculty, Saad Dahlab Blida University, PB 270, Soumaa road, 09000 Blida, Algeria
hemina.karim@etu.univ-blida.dz, {f_boumahdi,a_madani}@esi.dz

**Abstract.** Fake news is a term used to describe inaccurate information that has been disseminated to the public. This paper presents a novel approach to detect fake news using fine-tuned state-of-the-art language model, GPT-3. We fine-tuned the Generative Pre-trained Transformer 3 (GPT-3) model on the ISOT dataset to predict whether a given piece of text is likely to be fake news. We evaluated the performance of the model using accuracy, precision, recall, and F1-score. Our experiments showed that fine-tuning GPT-3 for fake news detection led to significant improvements in performance compared to the existing models. The fine-tuned model achieved an accuracy of 99.90%, a precision of 99.81%, a recall of 99.99%, and an F1 score of 99.90%. These results demonstrate the effectiveness of the fine-tuned GPT-3 model for fake news detection. Overall, our approach demonstrates the potential of fine-tuning GPT-3 for fake news detection, and the results suggest that it is a promising solution for addressing the growing problem of misinformation in social media and online news.

**Keywords:** Fake news · GPT-3 · deep learning · classification · cross validation

## 1 Introduction

Because they enable quick and free access to information, social media platforms have become vital sources of news for individuals in their everyday lives. While fake news is not a new phenomenon, it has become a big concern since, unlike traditional news sources such as newspapers and television, it is now easier to share misleading and fake news with a huge number of people via social media.

False news has had a substantial influence on events in several nations, including Brexit in Europe [1], the US presidential election 2016 [2], in Ethiopia, there have been fatal fighting between two ethnic groups, the Oromos and the Somalis [3], which was subject to fake news.


H. Zantout and H. Ragab Hassen (Eds.): ACS 2023, LNNS 760, pp. 41–48, 2023.
https://doi.org/10.1007/978-3-031-40598-3_5

The rest of this paper is structured as follows. In Sect. 2 we review notable existing work. Section 3 has a description of the used dataset. Section 4 is a description of GPT-3. In Sect. 5, we explain how we fine-tuned GPT-3. Section 6 contains the results achieved by our fine-tuned model. In Sect. 7, we explain how we validated our model using 5-fold cross-validation, and finally Sect. 8 concludes our work.

## 2 Related Work

Due to the serious issues caused by fake news, scholars have been working on its detection using several techniques and models. Some of them such as [4,5], and [6] worked on detecting fake news based on writing style. Mathews and Preethi [4] used Support Vector Classifier, while Nasir et al. [5] combined Convolutional Neural Networks (CNN) and Recurrent Neural Networks (RNN) to detect fake news according to writing style. Bidirectional Gated Recurrent Unit (Bi-GRU) was applied by Ghanem et al. [6].

Other scholars chose to fact-check the news by comparing it to an external set of previously known facts. Mayank et al. [7] build a knowledge graph fake news detection framework (DEAP-FAKES) for fake news identification using Graph Neural Networks (GNN). A combination of both fact-checking and writing style was used by Shemg et al. [8] in their framework Pref-FEND.

While both style-based detection and fact-checking use the textual content of the news, other researchers decided to detect fake news using contextual features. We can find Monti et al. [9] who trained their CNN model using the propagation tree of the news, and Matsumoto et al. [10] who decided to use GTN with user-related features such as profile introduction and the number of followers to detect fake news according to user credibility.

A combination of both content and context features can be used for fake news detection. Sharma and Kalra [11] used user characteristics with XGBoost and text characteristics with BERT transformer to detect fake news.

## 3 Dataset

Our model was trained on the ISOT[1] dataset [12]. It contains more than 21400 real news and more than 23400 fake news. the real news were obtained by crawling articles from Reuters, while fake news were collected from unreliable websites that were flagged by Politifact (A fact-checking organization in the USA). They decided to focus on political news articles separated into different types as shown in Table 1. For each article the following information is available:

- Article title
- Article text

[1] https://onlineacademiccommunity.uvic.ca/isot/2022/11/27/fake-news-detection-datasets.

- Article type
- Label (true or fake)
- The date the article was published on

**Table 1.** Breakdown of the categories and number of articles per category in the ISOT dataset

| News | Size (Number of articles) | Subjects | |
|---|---|---|---|
| Real News | 21417 | **Type** | **Articles size** |
| | | World news | 10145 |
| | | Politics news | 11272 |
| Fake News | 23481 | **Type** | **Articles size** |
| | | Government news | 1570 |
| | | Middle-east | 778 |
| | | US news | 783 |
| | | Left news | 4459 |
| | | politics | 6841 |
| | | News | 9050 |

## 4 Model Description

OpenAI's GPT-3 is a cutting-edge NLP model. GPT-3 is a deep learning model with 175 billion parameters that was pre-trained on a vast corpus of text data. It is supposed to create human-like prose by predicting the most likely next word or sequence of words given a certain context [13].

The model is capable of performing a wide range of NLP tasks, including language translation, text summarization, sentiment analysis, and question answering.

As shown in Fig. 1, GPT is composed of two main components [14]:

- **Multi-head attention:** Multi-head attention is an attention mechanism module that executes an attention mechanism numerous times in parallel. Their outputs are then concatenated and linearly translated into the expected dimension [15].
- **Feed Forward:** A feed forward artificial neural network that transfers information from the input layer to the output layer without using any feedback loops [16].

After each of the Multi-Head attention and the Feed Forward blocks, the input of the block is added to its output, and the result is normalized.

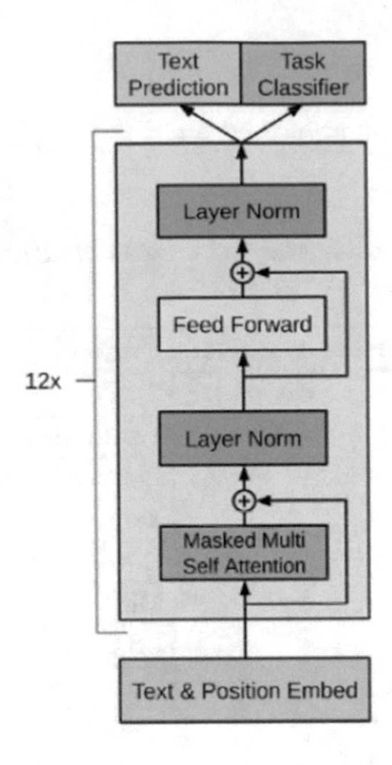


**Fig. 1.** Transformer architecture and training objectives of GPT [14].

## 5 Fine-Tuning

Given the encouraging results achieved by GPT-3 in language generation, and the high cost of pre-training such a model from scratch, we decided to fine-tune it for a classification task to detect fake news using the ISOT dataset. Our implementation follows the fine-tuning example released in OpenAI's documentation [17].

Three parameters were the focus of our parameterization.

- **Model:** OpenAI has four primary models for GPT-3 with varying levels of power that are ideal for various jobs [18]:
  - Davinci is the most capable model and can accomplish any task that the other models can, typically with less instruction and superior quality. It is most suited for applications that need a high level of comprehensions of the content, such as summarization for a specific audience and creative content generation.
  - Curie is both highly strong and extremely speedy. Although Davinci excels at understanding complex material, Curie excels at numerous delicate tasks such as sentiment classification and summarization.
  - Babbage is capable of simple tasks like classification.
  - Ada is typically the quickest model and can accomplish jobs such as text parsing, and some classification tasks that do not require a lot of content understanding.

We chose Ada for our model since it produced good results with text classification and is fast.

- **Number of epochs:** The number of epochs is a hyperparameter that determines the number of times the learning algorithm will run through the complete training dataset [19]. In our model, we decided to keep the default value of 4 epochs.
- **Batch size:** The batch size is the number of training examples used to train a single forward and backward pass, which is the number of examples to utilize before changing the model's internal parameters [19]. We fixed the batch size to 0.2% of the number of examples in the training set.

## 6 Results

To evaluate the performance of our fine-tuned GPT-3 model for fake news detection, as shown in Table 2 we compared it with several models that have been previously proposed in the literature.

We considered models that were trained on the same dataset as ours. The first model we considered is an SVC model trained by Mathews and Preethi [4]. The second model was proposed by Nasir et al. [5] and it is a combination of CNN and RNN. While the third model is a Bagging model proposed by Barbara et al. [20].

We also compared our fine-tuned GPT-3 model with two other models trained on the FakeNewsNet dataset, the first one is GNN model trained by Han et al. [21] and the second one is a combination of XGBoost and BERT proposed by Sharma and Kalra [11].

The results of our experiments showed that our fine-tuned GPT-3 model outperformed all benchmark models in terms of accuracy, precision, recall, and F1 score. Our fine-tuned GPT-3 model achieved an accuracy of 99.90%, a precision of 99.81%, a recall of 99.99%, and an F1 score of 99.90%, which was significantly higher than the performance of the benchmark models.

Our findings indicate the efficacy of our fine-tuned GPT-3 model for detecting false news and imply that fine-tuning GPT-3 can lead to considerable performance increases when compared to established benchmark models.

**Table 2.** Comparaison between our model and state-of-the-art models trained on benchmark datasets

| Citation | Model | Dataset | Accuracy | Precision | F1-score | Recall |
|---|---|---|---|---|---|---|
| Mathews and Preethi [4] | SVC | ISOT | 96.7 | 96.2 | 96.9 | 97.5 |
| Nasir et al. [5] | CNN + RNN | ISOT | 99 | – | – | – |
| Barbara et al. [20] | Bagging | ISOT | 99.64 | – | – | – |
| Han et al. [21] | GNN | FakeNewsNet | 85.3 | 83.4 | 84.1 | 85.2 |
| Sharma and Kalra [11] | XGBoost + BERT | FakeNewsNet | 98.53 | 98.20 | 98.57 | 98.51 |
| **Our fine-tuned model** | **GPT-3 ada** | **ISOT** | **99.90** | **99.81** | **99.90** | **99.99** |

## 7 Validation

In this paper, we used cross-validation as a standard technique to validate the performance of our fine-tuned GPT-3 model for fake news detection. Cross-validation is one of the most often used data re-sampling methods for estimating model true prediction error and tuning model parameters, it involves partitioning the dataset into training and testing sets multiple times and evaluating the model's performance on each partition [22].

We used 5-fold cross-validation approach in our experiments. Specifically, we partitioned the dataset into five equally sized subsets, with each subset containing an approximately equal number of real and fake news articles. For each fold of the cross-validation, we randomly selected four subsets for training and used the remaining subset for testing. We then fine-tuned the GPT-3 model on the training set and evaluated its performance on the testing set. We repeated this process five times, using a different subset for testing in each iteration. This approach helps to ensure that the results are reliable and generalizable to new data.

The results of the cross-validation showed that our fine-tuned GPT-3 model achieved consistent and high performance across all five folds. The average accuracy, precision, recall, and F1 score across the five folds were 99.98%, 99.99%, 99.88%, and 99.98%, respectively. These results demonstrate that our fine-tuned GPT-3 model is robust and reliable for fake news detection and can generalize well to new data.

## 8 Conclusion

In this paper, we presented a novel approach to fake news detection using fine-tuning of the GPT-3 model. Our experiments demonstrated that fine-tuning GPT-3 significantly improves the performance of the model for fake news detection, achieving high accuracy, precision, recall, and F1 score.

In addition to evaluating the performance of the model using standard metrics, we also conducted a cross-validation analysis to assess the consistency and generalizability of the model. The results of the cross-validation analysis showed that the fine-tuned model is consistent in its performance across different partitions of the data and can generalize well to new data.

Overall, the results of our study suggest that fine-tuning GPT-3 is a promising approach for addressing the problem of fake news detection. The model's high performance and generalizability make it a strong candidate for use in practical applications.

## References

1. Bastos, M.T., Mercea, D.: The Brexit botnet and user-generated hyperpartisan news. Soc. Sci. Comput. Rev. **37**(1), 38–54 (2019)
2. Bovet, A., Makse, H.A.: Influence of fake news in Twitter during the 2016 US presidential election. Nat. Commun. **10**(1), 7 (2019)
3. Dessie, B.A., Ali, A.C., Moges, M.A.: Towards ethno-political advocacy: Ethiopian journalists' professional role perceptions in post-EPRDF interregnum. Cogent Arts Human. **10**(1), 2168843 (2023)
4. Mathews, E.Z., Preethi, N.: Fake news detection: an effective content-based approach using machine learning techniques. In: 2022 International Conference on Computer Communication and Informatics (ICCCI), pp. 1–7 (2022)
5. Nasir, J.A., Khan, O.S., Varlamis, I.: Fake news detection: a hybrid CNN-RNN based deep learning approach. Int. J. Inf. Manag. Data Insights **1**(1), 100007 (2021)
6. Ghanem, B., Ponzetto, S.P., Rosso, P., Rangel, F.: FakeFlow: fake news detection by modeling the flow of affective information. In: Proceedings of the 16th Conference of the European Chapter of the Association for Computational Linguistics: Main Volume, pp. 679–689. Association for Computational Linguistics, April 2021
7. Mayank, M., Sharma, S., Sharma, R.: DEAP-FAKED: knowledge graph based approach for fake news detection. CoRR, abs/2107.10648 (2021)
8. Sheng, Q., Zhang, X., Cao, J., Zhong, L.: Integrating pattern-and fact-based fake news detection via model preference learning. In: Proceedings of the 30th ACM International Conference on Information & Knowledge Management, pp. 1640–1650 (2021)
9. Monti, F., Frasca, F., Eynard, D., Mannion, D., Bronstein, M.M.: Bronstein. Fake news detection on social media using geometric deep learning. CoRR, abs/1902.06673 (2019)
10. Matsumoto, H., Yoshida, S., Muneyasu, M.: Propagation-based fake news detection using graph neural networks with transformer. In: 2021 IEEE 10th Global Conference on Consumer Electronics (GCCE), pp. 19–20 (2021)
11. Shang, J., et al.: Investigating rumor news using agreement-aware search. CoRR, abs/1802.07398 (2018)
12. Ahmed, H., Traore, I., Saad, S.: Detection of online fake news using n-gram analysis and machine learning techniques. In: Traore, I., Woungang, I., Awad, A. (eds.) ISDDC 2017. LNCS, vol. 10618, pp. 127–138. Springer, Cham (2017). https://doi.org/10.1007/978-3-319-69155-8_9
13. Brown, T.B., et al.: Language models are few-shot learners. CoRR, abs/2005.14165 (2020)
14. Radford, A., Narasimhan, K.: Improving language understanding by generative pre-training. OpenAI (2018)
15. Vaswani, A., et al.: Attention is all you need, December (2017)
16. Bebis, G., Georgiopoulos, M.: Feed-forward neural networks. IEEE Potentials **13**(4), 27–31 (1994)
17. Open AI. Fine-tuning. https://platform.openai.com/docs/guides/fine-tuning. Accessed 16 Apr 2023
18. Open AI. GPT-3 documentation. https://platform.openai.com/docs/models/gpt-3. Accessed 24 Feb 2023
19. Brownlee, J.: What is the difference between a batch and an epoch in a neural network. Mach. Learn. Mastery **20** (2018)

20. Probierz, B., Stefański, P., Kozak, J.: Rapid detection of fake news based on machine learning methods. Procedia Comput. Sci. **192**, 2893–2902 (2021)
21. Han, Y., Karunasekera, S., Leckie, C.: Graph neural networks with continual learning for fake news detection from social media. CoRR, abs/2007.03316 (2020)
22. Berrar, D.: Cross-validation. In: Ranganathan, S., Gribskov, M., Nakai, K., Schönbach, C. (eds.) Encyclopedia of Bioinformatics and Computational Biology - Volume 1, pp. 542–545. Elsevier (2019)

# Enhancing Efficiency of Arabic Spam Filtering Based on Gradient Boosting Algorithm and Manual Hyperparameters Tuning

Marouane Kihal and Lamia Hamza(✉)

Laboratory of Medical Informatics (LIMED), Faculty of Exact Sciences, University of Bejaia, 06000 Bejaia, Algeria
{marouane.kihal,lamia.hamza}@univ-bejaia.dz

**Abstract.** Nowadays, the task of spam filtering in the Arabic language poses a unique challenge due to the complexity of its grammar and linguistic features, which makes it more challenging than in other languages. In this paper, we present a novel approach to enhance the efficiency of Arabic spam filtering using the Gradient Boosting algorithm and optimal hyperparameter selection. We examine the performance of our method using a greater range of experimental metrics than those used in the existing works. The experimental results demonstrate that our proposed model outperforms 6 Machine Learning algorithms for spam filtering in the Arabic language namely K-Nearest Neighbors, Decision Tree, AdaBoost, Naive Bayes, Logistic Regression and Support Vector Machines with an accuracy of 92.63%.

**Keywords:** Gradient Boosting · Arabic Spam · Machine Learning · Spam Detection

## 1 Introduction

Arabic is a Semitic language that is mainly used in the Arab world, which includes the Middle East and North Africa. With more than 420 million native speakers worldwide, it is included in the United Nation's six official languages. The Abjad script is the style used to write Arabic. It is written from right to left and comprises 28 letters, unlike the Latin alphabet, which is used to write several European languages the Abjad script uses many forms for each letter depending on where it appears in a word. The Arabic script has been utilized for long years, it was initially used to write the Arabic language, but it has since been adopted for writing Persian, Urdu, and Pashto. Arabic has a complicated grammatical system with numerous rules for word order, noun declension, and verb conjugation. The Arabic vocabulary is extensive and rich, with numerous terms coming from other languages, including Persian, Turkish, and Greek.


H. Zantout and H. Ragab Hassen (Eds.): ACS 2023, LNNS 760, pp. 49–56, 2023.
https://doi.org/10.1007/978-3-031-40598-3_6

Many websites and programs now support the Arabic language, making it simple for users to transition between languages. Arabic speakers may now access the internet and interact with online content more easily. Parallel to this, spam in Arabic, which refers to unwanted and unsolicited emails, texts, or other types of digital contact, is becoming a bigger issue online. It frequently contains viruses and other harmful software, links to malicious websites, and other elements that risk the security of computers and other electronic devices.

Numerous filtering techniques have been proposed to detect and prevent spam. These filters leverage diverse methods to identify and obstruct unwanted content, including keyword filtering, which entails scrutinizing content and contrasting it with a catalogue of established spam keywords. Machine Learning is another popular method, whereby Machine Learning algorithms are trained on a dataset comprising both spam and non-spam content to discern the distinguishing features of each type [1]. This trained algorithm can subsequently be utilized to classify newly-encountered text.

As, Gradient Boosting has gained popularity as an algorithm for spam filtering due to its capability to effectively handle high-dimensional feature spaces and imbalanced datasets. In this paper, we propose to use a Machine Learning algorithm called Gradient Boosting (XGBoost) to detect content spam efficiently. The proposed system is composed of three essential phases. First is content preprocessing, followed by extracting the important features. The Gradient Boosting algorithm is then employed to detect spam content by analyzing the extracted features, ultimately deciding whether the content is considered spam or not. Moreover, in our work we opted to find the optimal hyperparameters by making comparisons in different five evaluation metrics and changing these values to have a Gradient Boosting model more suitable to detect effectively the Arabic spam. The following are the paper's main contributions:

1. Proposal of a robust and efficient Arabic spam filtering system based on Gradient Boosting.
2. Experimenting with various hyperparameter combinations to develop an optimal Gradient Boosting model.
3. Comparing our approach against different Machine Learning algorithms using five evaluation metrics.
4. Generating a new dataset of Arabic spam messages.

This work is structured as follows: Sect. 2 presents the related works. Section 3 presents the architecture of our arabic spam filtering system. Section 4 presents the experiments and the results obtained. Section 5 concludes our paper and gives some perspectives for future work.

## 2 Related Works

The detection of spam has been extensively researched within the context of Email and web systems, and more recently in social networks [2]. Most of the works related to spam detection are based on Machine Learning [3–5]. Adel et al.

[6] proposed a bilingual Short Message Service (SMS) filter. This filter detects English and Arabic spam based on SMS content. The authors used Bayesian Naive on fourteen extracted features such as message length, spam words, presence of URL, Etc. In [7], suggested a system to address the issue of Twitter spam detection. The authors analyzed the content of tweets from Saudi Arabia to identify spam, and developed two variants: a rule-based approach and a supervised learning approach. The rule-based approach utilized a spam lexicon derived from the extracted tweets, while the supervised learning approach employed statistical methods based on the bag-of-words model and features identified through empirical analysis. In [8], the authors present a study on Twitter spam detection, focusing on classifying Arabic tweets as either automated (generated by bots) or manual. They proposed four categories of features, including formality, structural, tweet-specific, and temporal features. In terms of accuracy, their experimental evaluation showed that classifiers based on individual categories of features outperformed the baseline unigram-based classifier. In [9], the authors proposed an innovative filter designed to detect spam comments in social media networks operating within the Maghreb region and employing the Arabic language. The proposed filter utilizes Deep Learning, specifically Long Short-Term Memory (LSTM), and demonstrates superior performance to existing Machine Learning algorithms, as confirmed by the experimental evaluations. The authors in [10] presented a novel email spam filtering technique that utilizes Gradient Boosting on the "Spam Base" dataset [11]. The proposed method demonstrated better performance in identifying English spam compared to existing approaches, as evidenced by a diverse set of evaluation metrics employed beyond the conventional accuracy measure. In their study, Mussa et al. [12] introduced a gradient boosting filter that utilizes 15 features extracted from SMS messages, such as phone number, message length, and presence of spam words. Additionally, the authors conducted a comparison of their proposed approach with other boosting algorithms.

A comprehensive review of the existing literature showed that while many studies have been undertaken in the area of spam filtering for the English language [13,14], only a limited number have addressed the problem in the Arabic language. As such, our primary aim is to propose a more efficient system for Arabic content spam filtering.

## 3 Proposed Spam Filter

We started by cleaning and preprocessing the data to train our Arabic anti-spam filter. Then, a learning model based on the classification algorithms is built. This model is used in the test phase to decide whether the content is legitimate or spam, we specify that ham means no spam. Figure 1 illustrates the architecture of our Arabic spam filtering system.

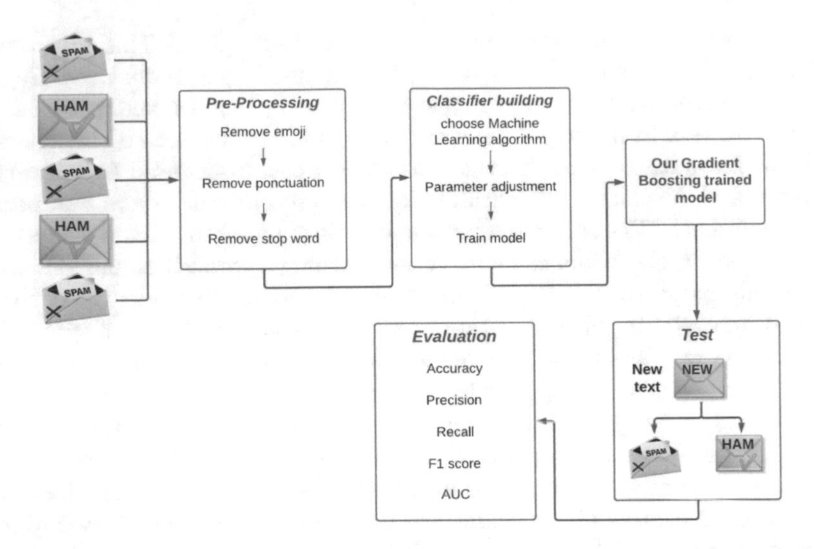


**Fig. 1.** The proposed Gradient Boosting spam filtering system

### 3.1 Pre-processing and Feature Extraction

During the cleaning and preprocessing phase, we first stripped emojis and punctuation from the text. Next, we utilized the Keras segmentation tool [15] to break down the words into tokens based on spaces. Afterward, each word was transformed into a vector using Tf-idf Vectorizer from Keras, which served as input for the classifiers. All test contents undergo this cleaning and preprocessing before classification.

### 3.2 Classification

Machine Learning is a set of techniques and algorithms used to train a machine to learn and make predictions based on training data. Some of these algorithms include: Support Vector Machines (SVM), Decision Tree (DT), K-Nearest Neighbors (KNN), Naive Bayes (NB), Logistic Regression (LR), and AdaBoost.

In this paper, as shown in Algorithm 1, we employ the use of Gradient Boosting (XGBoost), a highly effective supervised Machine Learning algorithm. XGBoost utilizes an ensemble approach by combining multiple decision trees in order to predict outcomes and mitigate overfitting, thereby enhancing the accuracy of the model. Gradient Boosting builds decision trees incrementally, with each tree being constructed to rectify the errors made by its predecessor. The trees are grown through the utilization of Gradient descent, which optimizes the loss function employed to assess the model's performance.

**Algorithm 1** Algorithm of arabic spam filtering using Optimal Hyperparameters and Gradient Boosting Algorithm

1- Collect an arabic dataset of labeled messages (1 for spam, and 0 for non spam).
2- Split the dataset into training and testing sets.
3- Train a decision tree on the training set as base model.
4- Calculate the residuals, which are the differences between the predicted values of the base model and the actual labels in the training set.
5- Train a new base model on the residuals calculated in step 4.
6- Repeat steps 4-5 for a set number of iterations
7- Combine the predictions of all the models into a final model
8- Use the final model to predict the message class
9- Evaluate the performance of the model on the testing set using accuracy, precision, recall, and F1-score.
10- Tune the hyperparameters of the Gradient Boosting algorithm, such as the learning rate, the criterion, and the maximum depth of the decision trees.

# 4 Experimentation

## 4.1 Dataset

For our experiments, we utilized the SMS Spam Collection [16], a publicly available dataset consisting of 5,574 SMS messages in English, of which 4,827 were non-spam messages and 747 were spam messages, all labeled as ham (non-spam) or spam. To generate a new dataset of Arabic spam messages, we manually translated each message from English to Arabic with the assistance of a native Arabic speaker, as part of our experimental process. However, publicly available Arabic spam datasets are limited in comparison to English datasets, which can make it difficult to find appropriate datasets for research purposes. Table 1 shows details of the dataset used.

**Table 1.** Statistics of dataset.

| Dataset | Spam messages | Ham messages | Total |
|---|---|---|---|
| The SMS Spam Collection [16] | 747 | 4827 | 5,574 |

## 4.2 Results

In our experiments, we used the Python programming language and specifically leveraged the scikit-learn library [17], which is a powerful tool for Machine Learning and data analysis in Python. This facilitated the efficient implementation and testing of a variety of Machine Learning algorithms on our dataset.

Manual hyperparameter tuning necessitates conducting manual experimentation with different sets of hyperparameters to determine an optimal combination of model parameters. We examine the performance of each combination using a greater range of experimental metrics than those used in the existing works namely Accuracy, Precision, Recall, F1-score and AUC.

Table 2 shows the result of a set of hyperparameters combination's. The most important hyperparameters that we have dealt with in training the Gradient Boosting model are:

1. Criterion: is the function to assess a split's quality. The terms "friedman mse" for the mean squared error with Friedman's improvement score and "squared error" for the mean squared error are used in our tests.
2. Max depth: is the number of nodes in the tree is limited by the maximum depth.

From the Table 2 we find that the combination friedman mse or squared error criterion with a max depth equal to 4 are the best combination. However, to better judge the efficiency of our model, we need to compare it to previous Machine Learning models. Table 3 shows that our xGBoost model outperforms 6 Machine Learning algorithms with an Accuracy of 92.63%, a Precision of 96.10%, a Recall of 95.48%, a F1-score of 95.79% and an AUC of 83.85%.

**Table 2.** Experimentation results.

| Criterion | max depth | Accuracy | Precision | Recall | F1-score | AUC |
|---|---|---|---|---|---|---|
| friedman-mse | 1 | 90.24% | 94.80% | 94.06% | 94.43% | 78.34% |
| friedman-mse | 2 | 91.83% | 95.97% | 94.74% | 95.35% | 82.32% |
| friedman-mse | 3 | 91.95% | 96.10% | 94.74% | 95.42% | 82.66% |
| **friedman-mse** | **4** | **92.63%** | **96.10%** | **95.48%** | **95.79%** | **83.85%** |
| friedman-mse | 5 | 91.60% | 95.97% | 94.49% | 95.23% | 81.90% |
| squared-error | 1 | 90.24% | 94.80% | 94.06% | 94.43% | 78.34% |
| squared-error | 2 | 92.06% | 96.10% | 94.87% | 95.48% | 82.87% |
| squared-error | 3 | 91.95% | 96.10% | 94.74% | 95.42% | 82.66% |
| **squared-error** | **4** | **92.63%** | **96.10%** | **95.48%** | **95.79%** | **83.85%** |
| squared-error | 5 | 91.60% | 95.97% | 94.49% | 95.23% | 81.90% |

**Table 3.** Results comparison.

| Classifier | Accuracy | Precision | Recall | F1-score | AUC |
|---|---|---|---|---|---|
| KNN | 87.75% | 93.50% | 92.54% | 93.01% | 72.46% |
| DT | 87.98% | 93.89% | 92.45% | 93.16% | 72.96% |
| AdaBoost | 89.45% | 95.83% | 92.35% | 94.06% | 77.35% |
| NB | 82.99% | 89.08% | 91.21% | 90.13% | 63.54% |
| LR | 85.14% | 91.29% | 91.64% | 91.47% | 66.94% |
| SVM | 89.13% | 96.10% | 91.69% | 93.84% | 78.40% |
| **Our xGBoost** | **92.63%** | **96.10%** | **95.48%** | **95.79%** | **83.85%** |

## 5 Conclusion

In conclusion, the problem of filtering Arabic spam is a challenging issue that can have severe consequences if left unaddressed. In this paper, we have explored the use of Gradient Boosting, a powerful Machine Learning technique, for classifying Arabic spam messages, and have demonstrated the impact of hyperparameters tuning on the performance of the model. Our findings reveal that certain hyperparameters, such as the criterion and the maximum depth of the trees, play a crucial role in achieving high accuracy. Our experiments show that our model outperform existing systems by achieving 92.63% of Accuracy. For future work, we plan exploring several areas to further improve the performance of Gradient Boosting, such as investigating more advanced feature engineering techniques.

**Acknowledgements.** This work has been sponsored by the General Directorate for Scientific Research and Technological Development, Ministry of Higher Education and Scientific Research (DGRSDT), Algeria.

## References

1. Saadi, D., Hamza, L.: Anti spam filter based on machine learning. In: International Conference on Advances in Communication Technology, Computing and Engineering CyberSpace, Morocco (ICACTCE 2021), March 2021. https://doi.org/10.26713/978-81-954166-0-8
2. Rao, S., Verma, A.K., Bhatia, T.: A review on social spam detection: challenges, open issues, and future directions. Expert Syst. Appl. **186**, 115742 (2021)
3. Kumar, N., Sonowal, S.: Email spam detection using machine learning algorithms. In: 2020 Second International Conference on Inventive Research in Computing Applications (ICIRCA), pp. 108–113. IEEE, July 2020
4. Zainab, A., Refaat, S., Bouhali, O.: Ensemble-based spam detection in smart home IoT devices time series data using machine learning techniques. Information **11**(7), 344 (2020)
5. Gupta, V., Mehta, A., Goel, A., Dixit, U., Pandey, A.C.: Spam detection using ensemble learning. In: Yadav, N., Yadav, A., Bansal, J., Deep, K., Kim, J. (eds.) Harmony Search and Nature Inspired Optimization Algorithms: Theory and Applications, ICHSA 2018, pp. 661–668. Springer, Singapore (2019). https://doi.org/10.1007/978-981-13-0761-4_63
6. Adel, H., Bayati, M.A.: Building bi-lingual anti-spam SMS filter. Int. J. New Technol. Res. **4**(1), 263147 (2018)
7. Al Twairesh, N., Al Tuwaijri, M., Al Moammar, A., Al Humoud, S.: Arabic spam detection in Twitter. In: The 2nd Workshop on Arabic Corpora and Processing Tools 2016 Theme: Social Media, p. 38, May 2016
8. Wahsheh, H.A., Al-Kabi, M.N.: Detecting Arabic web spam. In: The 5th International Conference on Information Technology, ICIT, vol. 11, May 2011
9. Kihal, M., Hamza, L.: Système Efficace de Filtrage des Spomments Maghrébins basé sur l'Apprentissage Profond Récurrent. Colloque International MOAD22, Méthodes et Outils dAide à la Décision, Université de Bejaia, Algérie, pp. 489–495, Novembre 2022

10. Mustapha, I.B., Hasan, S., Olatunji, S.O., Shamsuddin, S.M., Kazeem, A.: Effective email spam detection system using extreme gradient boosting (2020). arXiv preprint arXiv:2012.14430
11. Hopkins, M., et al.: Spam Base Dataset. Hewlett-Packard Labs, Palo Alto (1999)
12. Mussa, D.J., Jameel, N.G.M.: Relevant SMS spam feature selection using wrapper approach and XGBoost algorithm. Kurdistan J. Appl. Res. **4**(2), 110–120 (2019)
13. Ahmed, N., Amin, R., Aldabbas, H., Koundal, D., Alouffi, B., Shah, T.: Machine learning techniques for spam detection in email and IoT platforms: analysis and research challenges. Secur. Commun. Netw. **2022**, 1–19 (2022)
14. Guo, Y., Mustafaoglu, Z., Koundal, D.: Spam detection using bidirectional transformers and machine learning classifier algorithms. J. Comput. Cognit. Eng. **2**(1), 5–9 (2023)
15. Ketkar, N.: Introduction to Keras. Deep learning with Python, pp. 97–111. Apress, Berkelcy (2017)
16. Almeida, T.A., Gmez Hidalgo, J.M., Yamakami, A.: Contributions to the study of SMS spam filtering: new collection and results. In: Proceedings of the 2011 ACM Symposium on Document Engineering (DOCENG 2011), Mountain View, CA, USA (2011)
17. Pedregosa, F., et al.: Scikit-learn: machine learning in Python. J. Mach. Learn. Res. **12**, 2825–2830 (2011)

# Cyberbullying: A BERT Bi-LSTM Solution for Hate Speech Detection

Sihem Nouas(✉), Fatima Boumahdi, Amina Madani, and Fouaz Berrhail

University of Saad Dahleb, Somaa, Blida, Algeria
nouas.sihem.b3@gmail.com

**Abstract.** With the growth of data on social networks, the semantic analysis of shared text has become essential for decision-making and knowledge extraction. It is also the case for the detection of abusive and hateful content. Most of the work done in this field so far proposed classical machine learning solutions. But recently, models based on deep learning have also been applied. In this paper, we propose a neural network model for hate speech classification using two imbalance datasets that were imbalanced, containing approximately 25 thousand and 10 thousand annotated text samples. The proposed approach is evaluated using the standard evaluation metrics, such as Accuracy and F-score. We compared our approach to the baseline results and it showed satisfying results. Moreover, our approach was able to detect more instances of the minority class compared to the baseline results.

**Keywords:** hate speech · BERT · supervised learning · deep learning

## 1 Introduction

Hate speech has been growing considerably in recent years, In [9], the author stated "Hate speech is not about abstract ideas, such as political ideologies, faiths or beliefs - which ideas should not be conflated with specific groups that may subscribe to them. Hate speech concerns antagonism towards people ", the proportion of teens who say they often come across homophobic content has gone from 13% in 2018 to 21% in 2021, for sexist content the rate has gone from 15% to 21% and for racist content from 12% to 23% [14]. Various factors might contribute to this, and one of them is certainly the ability to anonymously post online [5] which contributes to the propagation of hate speech. The vast majority of the research done in this field proposed classical machine learning solutions, but in recent years, more researchers are using deep learning based approaches. In this paper, we have proposed a deep learning model to detect Hate Speech using two imbalanced Datasets (Twitter and stormfront forum) with approximately 25 thousand and 10 thousand annotated text samples. The rest of the paper is organized as follows: Sect. 2 presents some of the relevant work for hate speech detection. Our methodology and proposed Method is described in Sect. 3. Data set, experiments and results are discussed in Sect. 4; Finally, We conclude with the potential future scope of the study in Sect. 5.


H. Zantout and H. Ragab Hassen (Eds.): ACS 2023, LNNS 760, pp. 57–65, 2023.
https://doi.org/10.1007/978-3-031-40598-3_7

## 2 Related Work

Most of the previous work in this field mainly focused on lexical-based approaches [12,16] then, supervised classification using classical machine learning techniques, such as SVM [1,15,17], Random Forest Decision [2], and Logistic Regression [4,7,18]. The more recent solutions use neural networks models to automatically learn implicit representations using multiple layers from raw data [3,6,8].

The Lexicon-based approach usually uses pre-prepared sentiment lexicon to compute semantic orientation of a sentence or a document from the semantic scores of the lexicons. In [16] a Google bad word list was used as the hate lexicon to extract the features from the dataset, the count of hate words was manually extracted from each comment. In [12], the authors generated a lexicon of sentiment sentences using semantic and subjectivity features based on negative polarity, hateful words and theme-based Grammatical Patterns. A sentence was then classified to be in the hate category if it contained two or more words tagged as strongly negative. Otherwise, if it only contained a single strongly negative word the sentence was classified as weakly hateful. This method is straightforward and simple, and lexicon-based approachs might not take the context of the words around and thus cannot achieve high performances.

In [7], A multi-class classifier was trained to distinguish between hate speech, offensive language and neutral. After testing multiple models, the authors used Logistic regression with L2 regularization. Their model has achieved an overall precision of 91%, recall of 90% and F1-score of 90%. In [13], the authors used SVM to detect cyberbullying messages in English and Danish of Ask.fm datasets, because the datasets were imbalanced, the author applied a cost sensitive SVM using the LIBLINEAR library, the authors applied tokenization, PoS tagging and lemmatisation. Then, with different feature combinations, word n-grams, character n-grams and subjectivity lexicons using existing dictionaries. After optimization, the model achieved a maximum F1-scores of 58.72 % and 64.32 % for Dutch and English respectively. In [2], the authors performed a binary classification using a balanced dataset containing 520 tweet that were manually annotated into hate speech and not hate speech. When using word n-gram feature with Random Forest Decision Tree algorithm, the authors obtained an F-score of 93.5%.

In the last few years, various deep learning models with different architectures were introduced. In [11], the authors constructed a dataset from a white supremacy forum containing approximately 10 k sentences that were manually labeled (Hate, Non Hate). Their experiment was based on a balanced subset of the dataset summing up 2k labelled tweets. The authors implemented CNN, LSTM and SVM and reported that the LSTM classifier obtained the best results with 78% overall Accuracy. The authors in [6] built a convolutional neural networks (CNN) based model for hate speech detection using a total of 19905 for binary classification samples (Hate and non-Hate). The authors first pre-processed the data then used pre-trained word embeddings with the Google News Word2Vec model. The proposed model achieved an F-score of 80.35%. In [3], the authors also used a CNN-based model for cyberbullying detection using a dataset that contains 69874 tweets that were converted to Glove embedding vectors, their model achieved an overall accuracy of 93.97%. In [10], the authors used a dataset that contains 159k for binary classification (non-bullying and bullying) using three models, 1D-CNN model, LSTM and a Bi-LSTM that achieved an Accuracy of 96.33%, 94.12% and 97.45% respectively.

# 3 Proposed Approach

For hate speech detection, we use a Bi-LSTM based deep neural architecture with pretrained BERT embeddings using two datasets:

- Davidson's dataset [7]: it contains 24783 English tweets and has three classes, Hate: 1430 tweets, Offensive: 19190 and Neutral: 4163 Fig. 4.
- De Gibert's dataset [11]: it contains 10703 English sentences that have been extracted from Stormfront and has two classes, Not hate: 9507 samples, Hate: 1196 Fig. 5.

Our proposed method has been implemented using Keras, Tensorflow and Python.

## 3.1 Preprocessing

Text in natural language is usually unstructured and noisy. In our study, we preprocessed the text as follows:

- Convert all texts to lowercase.
- Replace some text contractions (example: "shouldn't" becomes "should not")
- Remove users, email, digits, Urls and re-tweets.
- Remove hashtag symbol from words.
- Stem words by replacing words by their stem (root).
- Tokenization.

## 3.2 BERT-Bi-LSTM

The proposed Deep Learning Architecture of hate speech detection is shown in Fig. 1. We have used pretrained BERT embeddings for English on Wikipedia and BooksCorpus[1] with a BERT preprocessor[2]. After preprocessing, the BERT encoder produces sequence outputs of size $128 \times 512$, these embeddings are passed into a self attention layer to focus on certain parts of the features, the output is then fed into a dense layer followed by a dropout layer ,another dense layer, a Bi-LSTM of 64 units, an additional Self Attention layer followed by a dropout layer, a flatten layer and finally a last dense layer with the number of class N that uses a Softmax function for the classification.

[1] https://tfhub.dev/tensorflow/small_bert/bert_en_uncased_L-6_H-512_A-8/2.
[2] https://tfhub.dev/tensorflow/bert_en_uncased_preprocess/3.

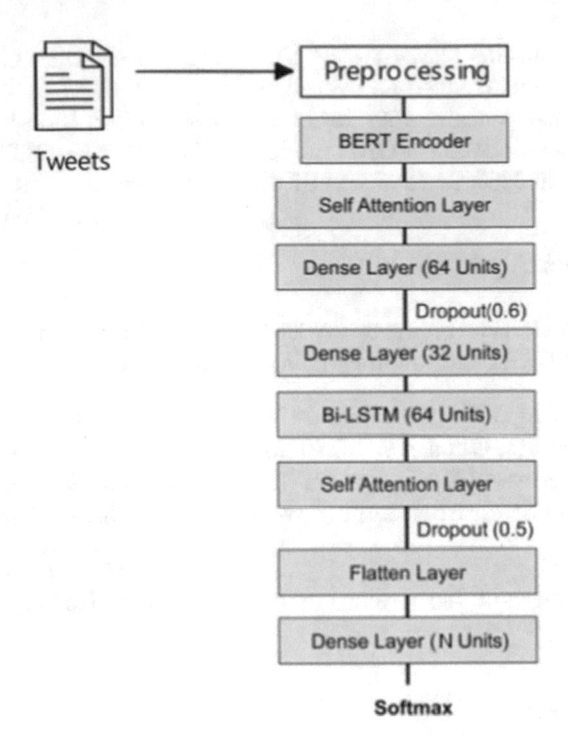


**Fig. 1.** Deep Learning Architecture for hate speech detection.

## 4 Experiment and Results

We used Davidson's dataset [7] and De Gibert's dataset [11] for hate speech detection. For both datasets, we split the data into 80% for training, 10% for validation and 10% for testing. Davidson's dataset is very imbalanced, 5.7% consist of hate tweets 77.4% of offensive tweets and 16.79% of neutral tweets. For De Gibert's dataset, only 11.2% of the sentences are labled as Hate speech. We avoid using oversampling or undersampling techniques since hate speech detection is a real life problem, instead, we run a second version of our neural network where we adjust the balanced weight for the cost function in order to give more attention to the minority classes. We compared the performance of our proposed method against the base line results.

In [7], the authors were able to reach an overall precision of 0.91, recall of 0.90, and F1 score of 0.90. Our model has reached an overall precision of 0.89, recall of 0.89, and F1 score of 0.89. However, when looking at Fig. 2, by using the weighted Bert-Bi-LSTM, the model was able to detect more instances of the Hate speech class. 0.43 of hate speech tweets were misclassified as offensive and 0.18 as neither, after testing the weighted model 0.12 were classified as neither and 0.15 as offensive tweets (Table 1).

**Table 1.** Results using the Pre-trained BERT BI-LSTM model comparison with the literature for each class.

| Dataset | Method | Class | Precision | Recall | F-score | Accuracy |
|---|---|---|---|---|---|---|
| [7] | Logistic regression | Hate | 0.44 | 0.59 | 0.51 | |
| | | Offensive | 0.96 | 0.91 | 0.93 | |
| | | Neutral | 0.83 | 0.94 | 0.88 | |
| | | Overall | 0.91 | 0.90 | 0.90 | 0.89 |
| | Bert-Bi-LSTM | Hate | 0.54 | 0.40 | 0.46 | |
| | | Offensive | 0.95 | 0.93 | **0.94** | |
| | | Neutral | 0.76 | 0.94 | 0.84 | |
| | | Overall | 0.89 | 0.89 | 0.89 | 0.89 |
| | Bert-Bi-LSTM (weighted) | Hate | 0.28 | **0.73** | 0.40 | |
| | | Offensive | 0.97 | 0.79 | 0.87 | |
| | | Neutral | 0.71 | 0.86 | 0.78 | |
| | | Overall | 0.88 | 0.80 | 0.83 | 0.80 |
| [11] | Bert-LSTM | Not Hate | 0.94 | 0.90 | 0.92 | |
| | | Hate | 0.42 | 0.58 | 0.48 | |
| | | Overall | 0.89 | 0.86 | 0.87 | 0.86 |
| | Bert-Bi-LSTM | Not Hate | 0.94 | 0.97 | 0.95 | |
| | | Hate | 0.67 | 0.46 | 0.54 | |
| | | Overall | **0.91** | **0.92** | **0.91** | **0.92** |
| | Bert-Bi-LSTM(weighted) | Not Hate | 0.95 | 0.92 | 0.94 | |
| | | Hate | 0.49 | **0.59** | 0.53 | |
| | | Overall | 0.90 | 0.89 | 0.89 | 0.89 |

We try to investigate the matter by looking into the tweets and the Word Frequency in both classes. Hate Tweets that were misclassified as offensive tend to have sexist and offensive language such as “RT @_2kkz: This fa**it bi**h tried to say she h*e'd for hd bearfaced bi**h yusa fa*” or “ni**as ain't playin a gram of defense man George need to tell these bi**h a** ni**as man up”. Same goes for Offensive tweets that were misclassified as hate speech like “Starks being a fa**ot” or “f***ing **eer”. As we can see in Table 2, these two classes share alot of common offensive words hence the misclassification Fig. 3.

In [11], the authors reported that an LSTM based model gave the best results with 0.78 accuracy. For comparaison we keep our architecture, but we use an LSTM layer instead of a BI-LSTM layer. We were able to outperform baseline results by achieving an overall accuracy 0.92, precision 0.91, recall of 0.92 and F1 score of 0.91. Identically to Davidson's dataset, the weighted model in De Gibert's dataset was able to detect more instances of hate speech from 0.46 in our base model to 0.59 correctly classified instances.

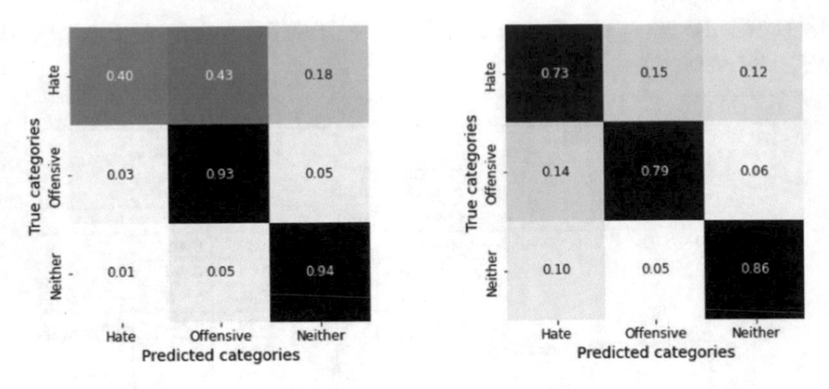


**Fig. 2.** Confusion matrix before and after weight balance for Davidson's dataset.

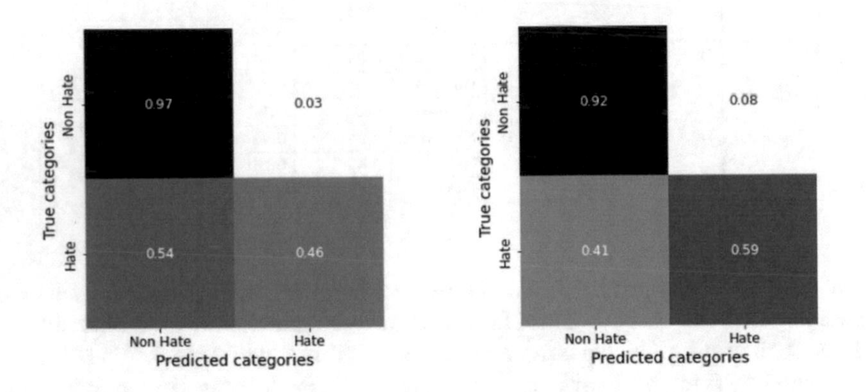


**Fig. 3.** Confusion matrix before and after weight balance for De Gibert's dataset.

| text | category | label |
|---|---|---|
| "@BabyAnimalPics: baby monkey bathtime http://... | neither | 2 |
| "@BaylaaGottaBody: 😂😂😂... | offensive_language | 1 |
| "@BlackChiquitita: Wow. RT @thatmanpalmer I'm ... | offensive_language | 1 |
| "@BlackNerdJade: Ok, sis." She'd rather be a b... | offensive_language | 1 |
| "@Blackman38Tide: @WhaleLookyHere @HowdyDowdy1... | hate_speech | 0 |

**Fig. 4.** Davidson's dataset with three classes (2: neither, 0: hate, 1: offensive).

| text | label |
|---|---|
| She may or may not be a Jew but she 's certain... | 1 |
| Thank you for the info. i will deffinetly look... | 0 |
| I believe this can be higher since I 've read ... | 0 |
| Thank you for posting your story . | 0 |

**Fig. 5.** De Gibert's dataset with two classes (1: hate, 0: Not Hate).

**Table 2.** Some of the most common words in each class for Davidson's Dataset.

| Class | Word | Frequency |
|---|---|---|
| Offensive | "b*tch" | 8927 |
| | "f*ck" | 1653 |
| | "ni**a" | 1474 |
| | "trash" | 281 |
| Hate | "b*tch" | 208 |
| | "f*ck" | 179 |
| | "ni**a" | 173 |
| | "trash" | 90 |
| Neutral | "trash" | 553 |

## 5 Conclusion

In this paper, we achieved satisfying results for hate speech detection outperforming baseline results with 0.92 Accuracy. We proposed a deep Bert bidirectional LSTM model with a Self Attention mechanism using two datasets. The minority class is more important in practical application, hence, we used a weighted version of our model which was able to detect more instances of the minority class that only constituted 5.7% and 11.2% in Davidson's and De Gibert's dataset with 40% to 73% and 46% to 59% (correctly classified instances respectively). Future research should consider using hybrid ensemble learning technique by implementing different weighted models in order to be able to detect more instances of the minority classes in the case of imbalanced datasets.

## References

1. Ahammed, S., Rahman, M., Niloy, M.H., Chowdhury, S.M.M.H.: Implementation of machine learning to detect hate speech in Bangla language. In: 2019 8th International Conference System Modeling and Advancement in Research Trends (SMART), pp. 317–320 (2019). https://doi.org/10.1109/SMART46866.2019.9117214
2. Alfina, I., Mulia, R., Fanany, M.I., Ekanata, Y.: Hate speech detection in the Indonesian language: a dataset and preliminary study. In: 2017 International Conference on Advanced Computer Science and Information Systems (ICACSIS), pp. 233–238 (2017). https://doi.org/10.1109/ICACSIS.2017.8355039
3. Banerjee, V., Telavane, J., Gaikwad, P.R., Vartak, P.: Detection of cyberbullying using deep neural network. In: 2019 5th International Conference on Advanced Computing & Communication Systems (ICACCS), pp. 604–607 (2019)
4. Br Ginting, P.S., Irawan, B., Setianingsih, C.: Hate speech detection on twitter using multinomial logistic regression classification method. In: 2019 IEEE International Conference on Internet of Things and Intelligence System (IoTaIS), pp. 105–111 (2019). https://doi.org/10.1109/IoTaIS47347.2019.8980379
5. Burnap, P., Williams, M.L.: Cyber hate speech on twitter: an application of machine classification and statistical modeling for policy and decision making (2015). https://doi.org/10.1002/poi3.85
6. Chaudhari, A., Parseja, A., Patyal, A.: CNN based hate-o-meter: a hate speech detecting tool. In: 2020 Third International Conference on Smart Systems and Inventive Technology (ICSSIT), pp. 940–944 (2020)
7. Davidson, T., Warmsley, D., Macy, M.W., Weber, I.: Automated hate speech detection and the problem of offensive language. In: ICWSM (2017)
8. Fang, Y., Yang, S., Zhao, B., Huang, C.: Cyberbullying detection in social networks using bi-GRU with self-attention mechanism. Information (Switzerland) **12**, 171 (2021). https://doi.org/10.3390/info12040171
9. Gagliardone, I.: Countering Online Hate Speech. UNESCO Series on Internet Freedom. United Nations Educational Scientific and Cultural, Paris, France (2015)
10. Ghosh, S., Chaki, A., Kudeshia, A.: Cyberbully detection using 1D-CNN and LSTM. In: Sabut, S.K., Ray, A.K., Pati, B., Acharya, U.R. (eds.) Proceedings of International Conference on Communication, Circuits, and Systems. LNEE, vol. 728, pp. 295–301. Springer, Singapore (2021). https://doi.org/10.1007/978-981-33-4866-0_37
11. de Gibert, O., Pérez, N., Pablos, A.G., Cuadros, M.: Hate speech dataset from a white supremacy forum. CoRR abs/1809.04444 (2018). URL: http://arxiv.org/abs/1809.04444
12. Gitari, N.D., Zhang, Z., Damien, H., Long, J.: A lexicon-based approach for hate speech detection (2015). https://doi.org/10.14257/ijmue.2015.10.4.21
13. Hee, C.V., et al.: Automatic detection of cyberbullying in social media text. PLoS ONE **13**, e0203794 (2018)
14. Rideout, V., Fox, S., Peebles, A., Robb, M.B.: Coping with covid-19: How young people use digital media to manage their mental health. San Francisco, CA: Common Sense Hopelab (2021). URL: https://www.commonsensemedia.org/sites/default/files/research/report/2021-coping-with-covid19-full-report.pdf
15. Rohmawati, U.A.N., Sihwi, S.W., Cahyani, D.E.: SEMAR: an interface for indonesian hate speech detection using machine learning. In: 2018 International Seminar on Research of Information Technology and Intelligent Systems (ISRITI), pp. 646 651 (2018). https://doi.org/10.1109/ISRITI.2018.8864484

16. Ruwandika, N., Weerasinghe, A.: Identification of hate speech in social media. In: 2018 18th International Conference on Advances in ICT for Emerging Regions (ICTer), pp. 273–278 (2018). https://doi.org/10.1109/ICTER.2018.8615517
17. Schuh, T., Dreiseitl, S.: Evaluating novel features for aggressive language detection. In: Karpov, A., Jokisch, O., Potapova, R. (eds.) SPECOM 2018. LNCS (LNAI), vol. 11096, pp. 585–595. Springer, Cham (2018). https://doi.org/10.1007/978-3-319-99579-3_60
18. Waseem, Z., Hovy, D.: Hateful symbols or hateful people? Predictive features for hate speech detection on twitter. In: NAACL (2016)

# Author Index





H. Zantout and H. Ragab Hassen (Eds.): ACS 2023, LNNS 760, p. 67, 2023.
https://doi.org/10.1007/978-3-031-40598-3

Printed in the United States
by Baker & Taylor Publisher Services